D1256963

The Parasites We Humans Harbor

THE PARASITES
WE HUMANS
HARBOR

by
Aaron E. Klein

ELSEVIER/NELSON BOOKS
New York

Acknowledgments

The author would like to thank his friend and teacher, Dr. Irving G. Kagan, Director, Parasitology Division, Bureau of Laboratories, at the Center for Disease Control. Dr. Kagan found time in an incredibly busy schedule to read the manuscript, help in the selection of pictures, and to spend many hours offering comments, suggestions, and criticisms. Thanks are also due to the staff of the Section of Comparative Medicine at the Yale University School of Medicine for allowing me to select pictures from their slide collection. Although the author is grateful for all the help he has received, he alone is responsible for the content of this book including any errors that may be in it.

Library of Congress Cataloging in Publication Data

Klein, Aaron E
  The parasites we humans harbor.
  Bibliography: p.
  Includes index.
  1. Medical parasitology–Popular works.
2. Parasitic diseases. 3. Host-parasite
  relationships. I. Title.
RC119.K57 1981   616.9'6   80–19727
ISBN 0-525-66693-1

Published in the United States by Elsevier/Nelson Books, a division of Elsevier-Dutton Publishing Company, Inc., New York. Published simultaneously in Don Mills, Ontario, by Thomas Nelson and Sons (Canada) Limited.

Printed in the U.S.A. First Edition

10 9 8 7 6 5 4 3 2 1

Contents

1.

The Parasitic Way of Life

So, naturalists observe, a flea
Hath smaller fleas that on him prey;
And these have smaller still to bite 'em;
And so proceed ad infinitum.
                    Jonathan Swift, *On Poetry. A Rhapsody*

The above is only one verse of a long poem written in 1733. Mr. Swift, the English writer and essayist who wrote, among other things, *Gulliver's Travels*, was not particularly concerned with fleas when he wrote his poem, since *On Poetry. A Rhapsody* was chiefly political satire. Nevertheless, Swift was expressing a basic fact of life, which is that practically every living thing on this earth, including you, the reader, has other living things residing in or on it. You are probably familiar with some examples of this very basic kind of living arrangement. If you have a dog or cat, you know that the animal might sometimes have problems with fleas living on it and worms living in it. What you are probably not as familiar with, or don't care to know about, is that there are thousands, if not millions, of living things subsisting in and on you. Most of your uninvited guests are bacteria, and few of them will ever cause you any pain or discomfort. You might also now have or have had or will have some creatures that could be described as worms living inside you. There is a better than 50 percent chance that there are some creatures domiciled in the skin around your nose that would probably make you gasp in horror if you saw one under a microscope. It is, however, unlikely that the "monster" (called the hair follicle mite, about which I will have more to say later) will ever cause you any harm.

The word that might come to mind when you think of all these organisms (an organism is a living thing) is "parasite." Indeed, some, but by no means all, of these organisms are parasites. Among the many words in our language, parasite is one of the most unpleasant. To call someone a parasite is to insult him, to imply that he gets something for nothing, does no work, takes and sponges from others without giving anything in return. This kind of parasite does not limit the takings to just money, food, and other tangibles, but can also freely take emotional support or depend on others to do all the mental and physical work of coping with day-to-day life. The term "social parasite" is often used to describe such people.

The word parasite did not always have such an unpleasant connotation. In ancient Greece a *parasitos* was somebody who got his meals in return for telling amusing stories at the dinner table. The biological concept of a parasite is somewhat different. One of the most difficult matters to tackle in any discussion of parasitology is, Just what is a parasite? I have made a start in answering this question with the idea of a living thing that lives in or on another living thing. Expanding that idea results in one of the most frequently offered definitions of a parasite—"an organism that obtains most or all of its needs from another organism while living in or on that organism." Sounds simple enough, but there is really much more to it than that. First of all, not all relationships that feature one organism living in or on another are parasitism. There are many different kinds of these living arrangements.

Scientists call "close living arrangements" *symbiosis*, and organisms that live in this way are called *symbiots* or *symbionts*. Symbiotic relationships range from those on a casual basis to those in which the symbionts' lives are so intertwined that one cannot live without the other. The benefits obtained or not obtained by the symbionts also vary widely. Some are mutually beneficial, but in others, one symbiont takes everything from the other and eventually causes its death.

Most biologists recognize four kinds of symbiotic relationships. These are *phoresis*, *commensalism*, *mutualism*, and *parasitism*. There are, however, no clear-cut divisions between them; it's more of an overlap, or blending. The first symbiotic relationship we will talk about is one in which one symbiont literally goes along for the ride.

Phoresis

Phoresis comes from a Greek word that means "to carry," and that is an apt description of what happens. The best-known example of phoresis is the relationship between many kinds of hermit crabs and sea anemones.

Unlike most crabs and lobsters, hermit crabs do not have hard shells protecting their bodies. To make up for this lack, they use the empty shells of dead sea snails. They carry these "mobile homes" with them wherever they go, retreating into them if they feel threatened. When a hermit crab outgrows the shell, it searches for a bigger one, moving into it as soon as one of suitable size is found.

Sea anemones are animals related to jellyfish. They look more like plants than animals. Most never, or hardly ever, move, spending their lives attached to rocks, the sea bottom, or other animals. They have thick round bodies, crowned by a ring of tentacles around the mouth. The tentacles, equipped with stingers, are used to capture and sting prey. The stinging paralyzes the prey so that it can be more easily swallowed.

In the phoresis that exists between some kinds of hermit crabs and sea anemones, the anemone is attached to the shell that is the "home" of the hermit crab. An anemone so attached is moved whenever the crab moves. Many of these moving arrangements are temporary, whereas others are more or less permanent, such as the phoresis between a species of hermit crab called *Pagurus* and the *Sagarita* anemone. Their phoresis starts when a *Pagarus* selects a shell that has a *Sagarita* attached to it. Once this selection is made, the two are lifelong companions. When the crab outgrows its shell and chooses another, it removes the anemone from the old shell and carefully places it on the new one.

Whether or not any of the partners in this arrangement receive any benefit is debatable. Both the crab and the anemone are capable of living alone and of taking care of themselves. The anemone might gain some advantage in finding food, and might also protect the crab from predators that are wary of the anemone's stingers, but there is no proof of this kind of benefit.

Another phoretic relationship exists between spider crabs

and small sponges. The sponges are attached to the crab and get a ride when the crab moves. It seems, however, that the crab does get something out of this arrangement. When the crabs are covered with these sponges, they are provided with protective camouflage. It is difficult to see any particular advantage for the sponge. Sponges don't really need to move. The seawater brings food to them when it circulates through the sponge, and the sponge filters out tiny organisms. This relationship goes beyond phoresis and is more like commensalism.

Commensalism

Commensalism means "eating at the same table." It is generally defined as a symbiotic relationship in which one partner benefits and the other is neither harmed nor benefited. However, there are many variations and borderline cases.

The classic case involves pilot fish and sharks. Pilot fish swim along with many species of sharks. The pilot fish stay close to the shark, without actually touching it, and they follow every change of speed and motion the shark makes. Sharks are not noted as particularly fastidious eaters, and there are always a few scraps around during and after a meal, which the pilot fish eat. The shark apparently gets nothing out of this relationship, but neither is it harmed. It was once thought that the pilot fish helped the poorly sighted sharks to find their way around—hence the name pilot fish—but sharks, which generally depend on smell more than vision, don't need pilot fish to find food for them. Pilot fish do not limit themselves to sharks, and often accompany other large fishes.

A somewhat closer relationship exists between sharks and a fish called the remora, which attaches itself to the shark with a sucker disk. The remora gets a free ride and scraps left over from the shark's meals. The shark does not seem to get anything out of this arrangement and, other than having to carry a little excess baggage, does not seem to be harmed. There is, however, some evidence that the remora eat tiny shrimp that cluster on the shark's body. I have seen sharks in aquariums behave as if they were irritated by the remora and were trying to shake it off. Remoras are not particular. They will attach to many different kinds of fish, and even to large sea turtles.

Another commensalistic relationship is that between the clown fish and the sea anemone (many different species). The fish hovers among the anemone's stinging tentacles and eats scraps of the anemone's meals. The clown fish, protected by a covering of mucus, is not harmed by the stinger cells. This protective covering develops only after the fish associates itself with the anemone, and so at first the fish does get some mild stings. If the fish leaves the anemone for a while (and they often do), the mucus coat thins out or disappears, but returns if the fish goes back to the anemone.

Mutualism

The strict definition of mutualism is a close relationship between two organisms in which both receive some benefit, but most biologists believe that for the relationship to be true mutualism, it must be a situation in which neither partner can live without the other. There are many examples of mutualism in nature. The best known is the lichen. Lichens are a mutualistic relationship between a fungus and a green algae, a kind of plant. (Some biologists argue that algae are not really plants, and that protozoa are not really animals. See page 191.) The algae provides food through photosynthesis, and the fungus supplies "housing." Neither the algae nor the fungus could live without the other. There are hundreds of different kinds of lichens, and they are found in almost all parts of the world. They are everywhere, but are easy to overlook. A common type occurs as a crusty growth on stones. Lichens are probably among the earliest forms of life to evolve on the earth.

Another well-known example of mutualism is the relationship between termites and one-celled organisms that live in the termite's intestine. The one-celled organisms are a type of animal called protozoa. It is a well-known fact that termites eat wood. What is not so well known is that termites can't digest the wood they eat. It's the protozoa in their intestines that do it for them. The mutual benefit is complete: The termites take in the wood and in so doing make the nutrition from the wood available to both partners in the arrangement. The dependence of the two organisms on each other is total; one couldn't live without the other.

There is some confusion about the words symbiosis and mutualism. Symbiosis was once used to describe relationships such as the one between the termite and the protozoan, but most biologists prefer the definition of symbiosis given on page 11. Mutualism is now used only to describe very close relationships in which the life functions of the partners are so intermixed that one cannot possibly survive without the other.

Cleaning Symbiosis

Among the more interesting partnerships are those called *cleaning symbioses.* In a cleaning symbiosis the partners do not constantly live in close contact with each other, nor are they dependent on each other for survival. Skin divers in warm tropical waters often see shrimplike animals walking over the bodies of fish. These shrimp remove parasites from the body surfaces of the fish. The shrimp get a meal and the fish get a cleaning. Small fish also have cleaning arrangements with larger fish. Some of these cleaning shrimp and fish work out of specific locations. Fish have actually been observed to wait in line, like patrons at a barber shop, for their turn to be cleaned.

One of these fish seems to be particularly brave or foolhardy. It swims into the mouth of a barracuda, a ferocious fish, to eat parasites that are in the barracuda's inner mouth surface and between its teeth.

These eating-cleaning relationships are not limited to the oceans. The Egyptian plover is a bird that picks parasites out from between the teeth of Nile River crocodiles. The tick bird cleans the hides of rhinoceroses, and egrets eat parasites from the backs of domestic cattle in Africa.

Looser relationships, somewhat like phoresis, exist between various birds and a number of large hoofed mammals, such as buffalo, in Africa. The birds sit on the backs of the mammals and dine on insects rustled out of the bush by the larger animals. Just what, if anything, the buffalo get out of this arrangement is not clear. There is some speculation that the birds serve to warn the buffalo of danger; that is, if the bird is startled by something, it will fly away, thus warning the larger animal of possible peril.

Parasitism

The general definition of parasitism offered before might bear repeating here: "an organism that obtains most or all of its needs from another organism while living in or on that organism." The organism in or on which the parasite is living is called, appropriately enough, the *host.* What should be added to this definition is that the host is the loser in this relationship. Hosts are usually harmed in some way. The harm varies from mere inconvenience to disease situations that can result in the death of the host.

Definitions aside, it is not always easy to determine whether a relationship is really parasitism. On the other hand, some organisms are clearly parasites. A tapeworm, for example, resides in the digestive tract of its host. Everything the tapeworm needs is obtained from the host. It doesn't even have a digestive system of its own, depending entirely on the host to do the digesting of food. The tapeworm just soaks up the digested food, somewhat like a sponge. There is no way an adult tapeworm can live outside the host in a natural situation. (Parasites have been kept alive out of hosts in laboratories with the use of elaborate, expensive equipment.)

It seems there is nothing good to be said about tapeworms, but, if nothing else, it can at least be said that there is no doubt or confusion about its status. It *is* a parasite, and a good example of one.

There is, however, doubt about the parasitic status of many organisms generally thought of as parasites. Insects called ticks, for example, feed on blood from warm-blooded hosts, such as dogs and, not infrequently, people. The tick stays on the host long enough to gorge itself on some of the host's blood. Female ticks increase in size enormously because the blood actually blows them up like a balloon. When they have had their fill, they drop off the host and can live for some time on the proceeds of one meal. A tick bite can produce a bothersome wound, open to infection. The bite of ticks can also transmit diseases such as Rocky Mountain spotted fever. Ticks stay on the host long enough, and inflict enough discomfort to the host, to be deserving of the name parasite.

Mosquitoes also take blood from a host while they are on the host, but they stay on the host for a very short time, often less than a second, to avoid being injured or killed by the host. A mosquito is on the fringe of parasitic life, if it can be considered a parasite at all.

Animals must eat to stay alive, and parasitism is just one way of getting food. An animal that lives by *predation*, by killing and eating other animals, is called a *predator*. Flesh-eating animals are also commonly called carnivores. *Carnivores*, strictly speaking, belong to the order Carnivora, and are meat-eating animals such as cats (big and small) and dogs. The term, however, is often incorrectly used in speaking of any animal that eats other animals. Insect-eating animals are called insectivores. Plant-eating animals are called *herbivores*. And those that eat both plants and animals are *omnivores*. People are omnivores.

Eating is a way of getting the energy that is needed to maintain life. The ultimate source of the energy needed by living things is the sun, and the only living things on this planet that are able to utilize the energy of the sun directly are organisms called *autotrophs*. Green plants are the best-known examples of autotrophs. They are green because they have within their cells a green substance called chlorophyll.

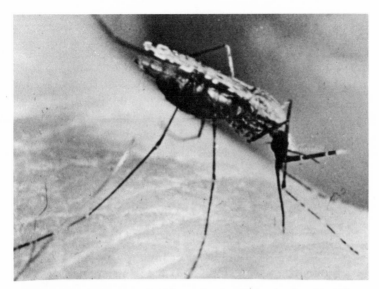

This mosquito, treading lightly on human skin, is about to take a blood meal. Whether it is a parasite or predator when it so sups is a debatable matter. This particular mosquito is called Anopheles gambiae. Anopheles *mosquitoes transmit the parasites that cause malaria.* **THE CENTER FOR DISEASE CONTROL**

When sunlight comes into contact with chlorophyll, a remarkable process takes place. This process, photosynthesis, results in the manufacture of food. The food made by plants is a basic carbohydrate molecule. From this carbohydrate, more complex carbohydrates are assembled.

An organism that cannot manufacture its own food is a *heterotroph*. Heterotrophs cannot derive energy directly from the sun, and so must eat plants and/or animals. All animals are heterotrophs.

When a herbivore eats a plant, the energy locked in the plant is passed on to the herbivore, and when a predator eats the herbivore, the energy is passed on again. If the predator is eaten by another predator, such as a carnivore, the basic energy is passed on yet again. This eating and being-eaten process is a food chain. It can be summarized in this way:

energy from sun⟶autotroph⟶herbivore⟶predator
larger predator⟶still larger predator, etc.

Of course, it isn't that simple. First of all, a food chain is hardly ever just a straight line of eating. Any given herbivore can feed on many different kinds of plants, and the herbivore in turn can be food for many different kinds of predators.

Another complicating factor is that green plants aren't as independent as they seem to be. They depend on animals for carbon dioxide, a basic raw material of photosynthesis. Animals exhale carbon dioxide during respiration. Although plants could exist by exchanging oxygen and carbon dioxide among themselves, there would not be the rich variety of plant life that exists on this planet were it not for the exchange of nutrients and gases among animals and plants. The energy given up by dead and decaying animals fertilizes the soil in which plants grow.

There are also organisms called *saprophytes.* These organisms live on the remains of dead organisms. Green plants, although dependent on soil fertilized through the products of animal decay, are not saprophytes. The difference is that green plants need certain chemicals from the soil that are produced when animals and plants decay. Saprophytes feed directly on the remains of dead organisms.

All fungi (molds, mushrooms, mildew, et cetera) are either saprophytes or parasites. In many instances, the once-living material on which saprophytes feed is pretty far removed from the living state. For example, bread mold lives on a slice of bread made of flour that was ground from wheat that once lived in a field.

All living things are dependent on each other in a relationship that is more like a web than a straight line. Where do parasites fit into this web? Parasites, by their very nature, are heterotrophs. That doesn't mean they are all animals. Bacteria, for example, are not animals, and neither are fungi; but most of them live all their lives as parasites or saprophytes. In some older books, you might find bacteria and fungi referred to as plants, but bacteria and fungi lack chlorophyll, so many biologists do not think they can rightfully be called plants. On the other hand, there are a few kinds of bacteria that are perhaps the only true autotrophs. These are the so-called iron and sulfur bacteria that get their energy from inorganic compounds of iron and sulfur. There are also parasites that are plants in every sense of the word except that they don't have cholorophyll, and they obtain their needs by living as parasites on other plants. An example is a plant called the dodder. It starts life as a "good citizen" with roots in the soil. However, as the dodder, which grows as a vine, wraps around the host plant, the dodder's roots wither away. Special absorbing structures grow into the host plant and take nutrients from it.

Animal parasites are very much like predators, but with some important differences. The relationship between a predator and its prey is a short one. The predator must kill its prey. A well-adapted parasite does not kill its host, at least not immediately. Some parasites do not take anything at all directly from the host; they eat some of the food that the host has eaten. Other parasites feed directly on the host, taking blood or other body tissues. However, although parasites can damage and weaken the host, even those that eat the host's tissues take but very little compared to a predator. A predator eats most, if not all, of the prey, and even if the predator is a fussy eater and eats only a portion of its kill, the remainder will be eaten by some other predator, a scavenger, or utilized by saprophytes. The parasite is actually far more considerate of the host than the predator is of its prey. Yet a predator, such as the carnivorous lion or tiger, is generally admired as a "noble beast," and a parasite is usually maligned as a sneak and a thief.

Parasites turn the table a bit in the food chain or web. The prevailing pattern in the food chain/web is that smaller animals are eaten by larger animals. Parasites are always smaller, much smaller than their hosts. Parasitizing, either inside the host or on it, is the only way a small heterotroph can prey successfully on a larger one. Although the lion might seem to be at the peak of its particular food chain as the "king of the jungle" (lions live in open plains, not jungles or rain forests), it really isn't. The worms living in its intestine and the thousands of other parasites lions have in and on them are the real "kings of the jungle."

The fact that some parasites live in the host and others on the host, and that some animals are more clearly parasites than others, suggests there is more than one way to be a parasite. That is, indeed, the case. Biologists, always eager to classify, recognize several kinds of parasitism. However, as with all systems of classification, there are no clear boundaries between the various ways of being a parasite.

Living in and living on are different ways of being a parasite. A parasite that lives in its host is an *endoparasite*, and a parasite that lives on the host is an *ectoparasite.* The tapeworm is a creature of clear status—definitely an endoparasite. Ticks are ectoparasites. However, a parasite that lives in the mouth of its host and occasionally wanders out of the mouth is difficult to classify. Similarly, certain types of flukes (relatives of tapeworms, *see* Chapter 6) spend their lives attached to either the gills of fish or on the surface of the fish. Those on the gills are endoparasites, whereas those on the skin are ectoparasites.

Some parasitologists will argue that even a tapeworm is not a true endoparasite. The tapeworm and many other parasites live in the interior space of the intestine called the *lumen.* The lumen of a garden hose, for example, is the part through which the water passes. The lumen of the intestine is the part through which the food, undigested and being digested, passes. The food material that passes through the small intestine is really not part of you until it is absorbed into the blood and assimilated into your cells. That food, although inside the human body, is really outside but within.

Tapeworms and other parasites that live in the lumen of the intestine are in the same relationship to you as the food moving through your intestine. They are inside you, but not part of you. There are parasites that really do live inside the host in every sense of the word. Some live inside individual cells, while others live in the blood and within other body tissues. The filarial worms (page 101) are examples of parasites that live within body tissues.

As one might expect, there are many in-between cases. Trichina (page 86) has stages that live in the lumen of the intestine and in the tissues. There are others, including trichina, that live so closely attached to the wall of the intestine

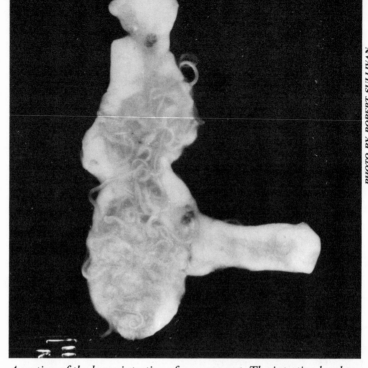

A section of the large intestine of a marmoset. The intestine has been cut open to reveal hundreds of parasitic roundworms.

YALE UNIVERSITY SCHOOL OF MEDICINE,
DEPARTMENT OF COMPARATIVE MEDICINE.

that they could be considered to be living within the body tissues.

An *obligatory* parasite is an organism that has to be a parasite if it expects to stay alive. A *facultative* parasite is one that can live alone, but can also live as a parasite under certain conditions. *Periodic* or *sporadic* parasites visit a host every now and then to fulfill some need and then leave.

A *hyperparasite* is a parasite that lives in or on another parasite. There are many examples of hyperparasitism, probably more than are now known. Fleas, for example, are parasitized by protozoans that live in their intestinal tracts. These are not the "lesser fleas" of which Jonathan Swift wrote in *On Poetry*. The idea, however, is the same. Some species of snails harbor stages of parasitic worms called flukes (page 159). Some species of the parasites in the snails are parasitized by organisms called *microsporidians*. And in a relationship more like that rhapsodized by Jonathan Swift, some ticks are parasitized by smaller ticks, which have yet smaller arthropods on them, and there are protozoans in the guts of all of them.

Parasitism, in all its many forms, is by no means an isolated or rare phenomenon on this planet. As a way of life, it is as commonplace as any other way of getting food. Practically all animals have parasites at some point in their lives. You might have seen movies in which there were magnificant scenes of massive herds of wild animals such as giraffes or antelope moving across the plains. However, only a very small fraction of all the animals in that scene are visible. What you don't see are the thousands of parasites in and on every one of these giraffes, antelope, deer, gnus, elephants, and whatever else is visible.

One might think that if all those animals have parasites, they must be sick, for it is well known that parasites cause diseases. Yet the healthiest looking wild or domestic animal is probably harboring parasites of one kind or another. Parasites do cause disease and, at any given time, an animal that seems to be doing well can fall victim to a disease brought on by the parasites that have always been there or by ones newly acquired.

Whether a host reacts to its parasites with indifference, annoyance, or illness is due to many factors. One of the most important is how many parasites the host has. One hook-worm, for example, will take about 0.5 milliliter of blood a day from its host. Considerably much more than that is lost from a minor shaving cut; the host will never miss the blood or know the parasite is there. However, if there are 500 hookworms, the host will lose about 250 milliliters of blood (about half a pint) a day. The host will then be very sick indeed.

Some parasites are more likely to make the host sick than others. The relationship between host and parasite is a two-way reaction: The host reacts to the parasite, and the parasite reacts to the host. The particular part of the host in which the parasite lives is the parasite's environment, just as the ocean near the shore is the environment of certain species of crabs, and the desert is the environment of the horned toad.

Over millions and millions of years, organisms have adapted to their environments, they have developed ways of surviving. Parasitic organisms are no exception. Since para-sitism implies a two-way relationship, the host has also be-come adapted to the parasite.

One extreme in a host/parasite relationship is that a par-ticular parasite cannot possibly survive in a particular host. The other extreme is that a particular parasite causes the death of the host. At the middle is an "ideal" relationship in which the parasite gets what it needs from the host while doing the host minimal harm. This balance between host and parasite can change, which is why a healthy animal can become sick from the parasites it has always had.

Some hosts can develop an immunity to a particular kind of parasite. The first time the host becomes infected, its body responds by producing substances that slow down the activities of the parasite or even kill it. In subsequent invasions by that particular parasite, the parasites will have a harder time or be unable to become established in the host.

In the next chapter we will take a look at some of the ways parasites have become adapted not only to survive as individuals but to survive as a species—that is, reproduce and spread from one host to another.

2.

That Thing Might Be Living in *Me?*

The parasitic way of life seems to be one filled with advantages and notably devoid of disadvantages. Although many species of parasites get all they need with only a fraction of the work free-living animals have to do, the parasitic way of life is not without problems. The biggest problem of parasites could be stated as "What are we going to do with the children? They can't go on living with us forever."

Finding a new host for the offspring is probably the largest problem a parasite has. The parasite is locked into the host. If the species is to survive, there must be a way to get the offspring out of the old host and keep them alive long enough in the outside world to find a new host. The prevailing pattern in parasitism is that offspring, usually in the form of eggs, leave the old host and seek another, rather than stay. If they stayed, the host would soon become overloaded with parasites and would probably die as a result. If that happens, the parasites will die, too.

Parasites have developed some remarkable ways of increasing the chances that their offspring will find a new host, but before we say anything more about it, some warnings are in order. Parasites are not the sort of organisms that "think" or "plan" anything. They are incapable of thinking about anything. All actions of parasites, as ingenious as they might appear to be, are automatic, preprogrammed responses.

Another word for preprogrammed responses is *instinct.* Instinct is a pattern of behavior the animal inherits from its parents. It doesn't have to learn anything. Everything the organism needs to do in order to survive is done through instinct. The instinctive patterns of behavior are as necessary for the animal's survival as any physical characteristic, such as the ability to move quickly.

We are animals, too, but we have very few instinctive behavior patterns. We learn what we have to know to cope and survive, and, for humans, the learning process is very long; indeed, it never ends. Other animals seem to have both learned and instinctive behavior patterns. A dog, for example, can learn how to do many things, but it is also born with many instinctive behavior patterns.

An animal's total package of instincts, physical characteristics, ability to learn, and whatever else it has to help it survive are its *adaptations.* Examples of adaptations are a duck's webbed feet and the instinctive behavior pattern of oiling its feathers with oil secreted by its anal glands. A parasite's adaptations might include body chemistry that enables it to live with the rather limited oxygen supply in a host's intestines, and the behavior patterns that increase the chances of the offspring finding a new host.

An animal's adaptations are inherited from its parents, which in turn inherited them from their parents, and so on. All the adaptations of all the living things on the earth today are the product of billions of years of evolution.

Evolution is basic to all life, but it is widely misunderstood. That evolution is a process of change is fairly well known. However, just how these changes occur and how they are passed on from one generation of living things to another is not so obvious. Even Charles Darwin, who formulated the evolution concept (or the theory of natural selection) didn't know much about evolution.

Darwin lacked knowledge of genetics, which is how characteristics are passed on from parents to offspring. What we inherit from our parents is in the form of a package of materials called *genes.* When we start life as one fertilized egg cell—the product of the fusion of a sperm cell and an egg cell—we have all the genes we are ever going to have. These genes determine what our set of adaptations will be. If you are an intelligent animal, such as a human, what you do with your set of adaptations is largely a matter of your own choice. If you are a tapeworm, there are no choices involved whatsoever.

An animal's set of genes are similar to those of its parents but not exactly like those of the parents. There is a mixing process in the reproductive organs that produce sperm and egg cells that serves to shuffle the genes and deal them out randomly to offspring.

Another factor involved in genetics is *mutation*, a spontaneous change that takes place in a gene. Suppose that in a group of newborn animals, such as a nest of fledgling birds, there is one bird that has inherited a mutated gene. This gene has made the bird's beak a little bit longer than the beaks of its nest mates or the beaks of all the other birds of the same kind. If these birds feed by picking insects out of tree bark, this mutated bird now has a competitive advantage over all the other birds of the same kind. It is more likely to be able to get at food in tree bark, and it has a better chance of staying alive long enough to reproduce and pass on the longer beak characteristic to its offspring.

Note that no bird ever thought to itself, "A long beak would be better for getting the bugs out of that tree, so I think I'll develop one." Evolution is a blind process. A mutation could just as easily be unfavorable to an organism's survival as it could be favorable. In all probability, most mutations that occur are unfavorable.

For any mutation to matter at all in evolution, it must occur in a cell destined to become a sperm or egg cell, and that cell must fertilize or be fertilized and become a new organism. Contrary to what might be seen in 1950's-vintage science-fiction movies on late-night TV, whole animals do not mutate to become raging monsters that tear down buildings and swat airplanes down from the sky. What mutates is a gene, specifically a section of a molecule of DNA, the substance of which genes are made. Mutations and evolution were happening on this planet even before there was what we know as life.

Mutations of nonliving material into substances that could be called living started life on this planet. One of the most widely quoted theories of the origin of life is that life started more or less spontaneously. The world was covered in an ocean, or "primeval soup," filled with certain types of molecules. Through energy provided by lightning or ultraviolet radiation (there was more ultraviolet radiation reaching the earth from the sun then because there was no ozone layer to filter it out), these molecules came together to make larger molecules with some very special qualities. These molecules could grow through taking on the surrounding molecules, and they could duplicate themselves—that is, reproduce. They were alive.

When life first started on this planet, there was more than enough food for all living things. The early, primitive forms of life did not have to do much to get what they needed. They just floated around, and the needed molecules used for food came to them. As millions of years passed, the early life became more complex, perhaps somewhat like today's bacteria and other one-celled organisms. As the population increased, the available food supply decreased, and it became harder to get what was needed. Those organisms that through evolution became better adapted to get food were the ones that survived. The world itself continued to change, and these changes caused some species to die out and others to prosper. Those that became adapted to the changing world survived; those that did not became extinct.

Biologists say that those organisms that just floated around and more or less soaked up food "lacked competence." The word "competence" as used here does not mean exactly the same as it does when used to describe how well some person can do a job. Rather, it refers to the animal's adaptations. As food became more difficult to procure, those organisms that developed a competent set of adaptations were able to get the food they needed.

Eventually, the two basic ways to get food, autotrophic and heterotrophic, evolved. The emergence of green plants that produced oxygen changed the atmosphere to what it is today. Parasitism also evolved. No one knows exactly how, but you can be certain that no animal thought to itself, "It's getting pretty tough to get anything to eat around here, so

I think I'll become a parasite and let the host do all the work." Parasitism evolved as blindly as anything else evolved.

We can make a few suppositions. Suppose, some hundreds of millions of years ago, some wormlike animals were swallowed by a fish. Most of the worms provided a welcome meal for the fish, but perhaps one of them somehow got lodged in the fish's mouth. This worm was a little different from the others. Its body chemistry enabled it to survive in the fish's mouth. The worm reproduced, and its offspring inherited the body chemistry that made it possible to stay alive in the fish's mouth and benefit from it. Some of the worms stayed in that fish, and others ventured out, some finding another fish to live in, and others living free or perishing. In time, perhaps the evolution proceeded to the point where the worms were so adapted that they could not survive outside the fish, and they had become obligatory parasites.

Parasitic relationships that exist today range from simple arrangements, such as a flea living on a dog, to complex relationships. Among the more complex behavior patterns that parasites have are those that involve getting into another host. Some are so complicated that it seems a wonder that such parasites ever find a new host at all. Some parasites seem to have taken advantage of the host's life-style. Others gain entry by making use of other parasites the host might have. But whatever the parasite's behavior pattern, it is not the result of any carefully thought-out plan. "Ingenious" would be a poor word to describe it; "lucky" is more appropriate. The behavior patterns are the results of millions of years of evolution. Those parasitic organisms whose mutations happened to take them in a path that was complementary to the host's evolution, so that entry into the host was facilitated, are the ones that survive today.

Most contemporary biologists agree that all parasites evolved from free-living organisms. As the organisms evolved into parasites, many of them reverted to the non-competence of the early days of life on this planet. Most of them encounter no competition for food inside the host. They just soak it up, as did the very earliest big molecules that could be considered to be alive.

The series of events in a parasite's life, from adult to production of offspring to the offspring becoming established in a new host, is the *life cycle.* There are many kinds of life cycles, some simple, some marvelously intricate and complicated. Practically all parasite life cycles, however, have one thing in common, and that is that enormous quantities of eggs are produced. The chances that any one of these eggs will ever become an adult parasite established in a host are very, very slim. The capacity of some kinds of parasites to produce hundreds of millions of eggs in their lifetimes helps to overcome these odds to some extent. Others have ways of getting the maximum out of every egg. In some species of parasites, one egg has the potential of giving rise to hundreds of thousands of new organisms.

Types of Life Cycles

The *Ascaris* worm has a comparatively simple life cycle. The adult worms live in the intestines of hosts such as pigs, horses, or people. The females lay eggs that pass out of the host in the host's intestinal waste products. A female *Ascaris* worm can produce about 200,000 eggs a day. Where there are no toilets, or where toilets are not used to any great extent, the eggs find their way to the ground. Under the right conditions, the eggs can survive for as long as five years. In order for the embryo worms in the eggs to become adults in another host, the eggs must be swallowed in contaminated food and on dirty hands placed in mouths. The eggs hatch in the intestines of the new host, after which the young worms go on a tour of the body, eventually resettling in the host's intestines. The *Ascaris* life cycle is an example of a direct life cycle: *host ⟶ host.*

The hookworm's life cycle is similar to that of the *Ascaris*, with the exception that the hookworm gains entry by burrowing directly through the skin into a surface blood vessel. The blood carries the embryo through the host's body and, eventually, into the small intestine where the worms make their home.

Many parasites employ one or more *intermediate hosts* before becoming established in the *main host*, or, as that host is called in this type of life cycle, the *primary host.* An example of a parasite with this kind of life cycle is a tapeworm that commonly infects dogs. Eggs pass out in the feces (waste products) of the animal. The eggs are then ingested by the larvae of fleas, common inhabitants of all parts of a dog's skin surface. As the larva matures into a flea, the egg hatches into an immature form of tapeworm. Anyone who has a dog knows that the dog is likely to give some vigorous attention to the site of a flea bite with teeth and/or paws. If the dog scratches at the bite with its teeth, it might swallow the flea. By so doing, the dog could well ingest immature tapeworms that will grow into adult tapeworms in the dog. This life cycle can be summarized as: dog-flea-dog, or, primary host–intermediate host–primary host.

Many parasites gain entry by getting in the host's food. The three most common tapeworms of humans are called fish tapeworm, pork tapeworm, and beef tapeworm because of the foods they are found in.

The fish tapeworm, for example, goes through development in at least two intermediate hosts before reaching a primary host such as a bear, dog, or human. Another dog tapeworm is ingested in the meat of rabbits, and a fairly common tapeworm of cats gains entry in the flesh of mice. Trichina, a relative of *Ascaris* and hookworm, is one of the most widespread of human parasites. It can get into a human host when pork, not cooked long enough to kill the larval worms in the meat, is eaten.

Intermediate hosts often help parasites to get at their primary hosts, although in some instances, it would seem that the complexity of intermediate host life cycles would serve to keep the parasites away from their hosts. Intermediate hosts also increase the reproductive potential of some parasites. Most species of a dangerous variety of parasite called blood flukes use snails as the intermediate host. The adult worms live in many places in the primary host, including the blood vessels near the digestive tract. Eggs leave the host in waste products and hatch if they are in water. The larvae that hatch from the eggs enter snails by burrowing directly into the snail's body. In the snail, the larva changes into another form that produces still another kind of larval form capable of infecting another primary host, such as a human, by burrowing directly into the skin. The form of the parasite that lives in the snail continues to produce larvae for quite some time, ranging from days to years depending on a number of factors, including how long the snail lives. However, one egg

LIFE CYCLE of—

Schistosomes

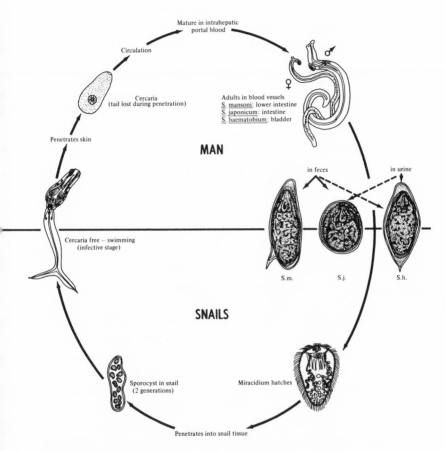

Mature in intrahepatic
portal blood

Circulation

Cercaria
(tail lost during penetration)

Penetrates skin

Adults in blood vessels
<u>S. mansoni</u>: lower intestine
<u>S. japonicum</u>: intestine
<u>S. haematobium</u>: bladder

MAN

in feces   in urine

Cercaria free – swimming
(infective stage)

S.m.   S.j.   S.h.

SNAILS

Sporocyst in snail
(2 generations)

Miracidium hatches

Penetrates into snail tissue

*A diagrammatic representation of the life cycle of schistosomes,
blood flukes. These parasites utilize snails as an intermediate host.
Inside the snail the parasites produce enormous numbers of new
forms capable of reinfecting a new main host—man.*

THE CENTER FOR DISEASE CONTROL

producing 200,000 or more infective larvae is not uncommon in the lifetime of the snail.

Many insects are carriers, or *vectors*, of parasites. We have already talked about the flea, which has to be eaten by a dog before it can convey a tapeworm into the dog's interior. Other insects introduce parasites into hosts by doing what insects do—bite. "Pierce" is a term that more accurately describes what happens when many insects bite. A mosquito, for example, pierces your skin (or the skin of some other warm-blooded animal) with mouth parts that are like a hypodermic needle. Before the mosquito draws out her blood meal (only the females bite), she shoots into the site a liquid from her salivary glands. This liquid keeps the blood from coagulating in her mouth parts and clogging them up. The salivary-gland liquid can harbor parasites. The parasites that cause malaria get into the primary host in this way. Some of the more horrible parasitic diseases are carried by insect vectors. Long, thin roundworms called filarial worms can block the lymph channels in the primary host (such as man), causing legs, arms, and other body parts to swell into grotesque shapes. This stage of the disease is called elephantiasis, an allusion to legs that look somewhat like elephant legs. A variety of biting flies in tropical areas carry parasites that can cause blindness and severe destruction of body tissues; others carry parasites that cause chronic, debilitating diseases. Charles Darwin probably contracted one of the these insect-borne diseases in South America.

Some parasites can live in only one particular host, while others thrive in a wide variety of hosts. Just why a parasite is able to live in one host and dies in another is not completely understood, but the body chemistry of both parasite and host is undoubtedly the main factor. The exact details of this chemical interaction between parasites and hosts are among the most widely studied aspects of parasitology today.

Another remarkable thing about parasites is that they are not digested or otherwise dispatched in the host. A parasitic worm living in the small intestine of its host is, like any other living thing, largely protein. Any protein in the small intestine is usually quickly broken down into its constituent parts through the action of digestive enzymes. It is obvious that parasitic worms are not digested.

Not being digested in the host's intestine is one of the ways the parasite is adapted to the parasitic way of life. The exact biochemical basis of this adaptation is the subject of a great deal of research. It is thought that the parasites might give off substances that counteract the action of the host's digestive enzymes. It is known that when they die, intestinal parasites are often digested by the host.

Parasitology, as a science, started around the middle of the last century, and it continues to grow. There are two basic ways to approach the study of parasitology. One is biological, and the other is medical. From a strictly biological viewpoint, parasites are interesting organisms that deserve to be studied for their own sake. From a strictly medical point of view, parasites cause some of the most terrible diseases of humanity and of the animals and plants that people depend on. Parasites must be studied so that more ways can be found to control them and to cure the diseases they cause.

Of course, the two approaches overlap. Understanding the biology of parasites is the first step toward controlling them. If a parasite causes disease in its host, there is generally some kind of imbalance present since the host is the parasite's environment. The imbalance could be an overpopulation of parasites in one host. We have heard much about the great harm that might be done to the Earth by overpopulation. The host is the parasite's world, and there is no doubt that too many parasites in one host is destructive to the host and often to the parasite. Another imbalance is one in which the host and parasite are not well adapted to each other. For example, the host might have allergic reactions to the waste products of the parasite.

The balance between parasite and host can change. Quite often, a host can harbor a few parasites and never show any ill effects. However, should the condition of the host change due to malnutrition or some other problem, the parasites could cause greater harm to the host.

History of Parasitology

Hundreds of years ago, when pioneer anatomists began to dissect human bodies, they often found parasites in the cadavers. They didn't know quite what to make of these wormy things. The prevailing idea was that they just grew there. This idea was widespread and not limited to the parasites. Until a little more than a hundred years ago, there were still many biologists who thought that some kinds of life could spring forth spontaneously. Maggots, for example, were thought to be formed from rotting meat, and mosquitoes supposedly sprang forth from water without benefit of parents. This theory was called *spontaneous generation.* In medieval times there were even books of recipes for making mice, frogs, and many other kinds of life.

The first observers of internal parasites did, of course,

wonder how the parasites got there. It is quite understandable that the idea that tapeworms, for example, sprang forth from the "foulness with the gut" was widely believed. Many of the early scientists thought that intestinal worms were beneficial. In London, one T. Moufet wrote in 1634, "Some putrefied, superfluous and fecal matter in us is evidently collected, the hand of benevolent nature turns it into worms and in this way purifies the body."

People have known about parasites, particularly parasitic worms, for a long time. There are references to them in the writings of ancient Egyptians. The Jewish dietary laws spelled out in Deuteronomy were motivated by the knowledge that you could get worms from eating the meat of certain animals, such as pigs and camels. Aristotle (384–322 B.C.) said there were three kinds of intestinal worms, ". . . large and flat, cylindrical, and thin." He was probably referring to tapeworms, ascarids, and pinworms. Pliny the Elder (A.D. 23–79), a Roman scholar, wrote a thirty-seven-volume *Natural History* in which he said that man had two kinds of worms, round and flat. Galen (A.D. 130–200), the great Greek physician who practiced in Rome, recognized three kinds of worms in people.

Although the existence of internal parasites, such as worms, was known for thousands of years, scientists did not begin to understand the life cycles of these organisms until the middle of the 1800's, partly because the idea of spontaneous generation stubbornly stayed in the minds of men. Not until Louis Pasteur finally disproved the concept of spontaneous generation in 1865 did the idea begin to fade. Even after spontaneous generation was no longer accepted, it took a long time to work out the remote relationships between a

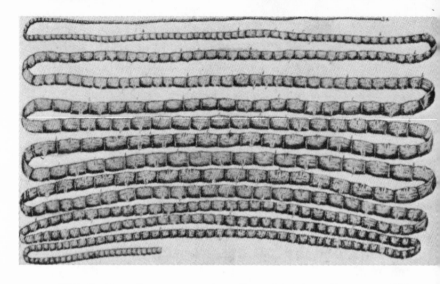

This drawing of a tapeworm dates from the eighteenth century. The gluttonous Louis XIV, king of France from 1643 to 1715, was successfully relieved of a large tapeworm similar to this one. Both the tapeworm pictured here and the king's companion were probably beef tapeworms.

tapeworm in a human gut and the so-called "bladder worms" in the flesh of cows, and that debilitating blood flukes came from placid, slow-moving snails.

Figuring out the life cycles of parasites resulted in some of the most fascinating detective stories in the history of science. A great deal of intuition, ingenuity, and just plain hard work was needed to work out some of the complex life cycles. While the biochemical relationships between host and parasite occupy the attention of many parasitologists today, there are still many life cycles that have not been worked out.

There is still much about parasites that is not known. Many young people interested in a career in the life sciences have found parasitology to be an interesting field, and one in which there are many opportunities.

You Have to Call Them Something

Every organism must have a name to distinguish it from all other living things. Names of animals are not usually a problem for us. From the time we are old enough to sit on a parent's lap we are exposed to pictures of animals in alphabet books and in stories and fairy tales. Such stories teach us to recognize animals. For example, most of us know a bear when we see one. However, although we call it a bear, the *common name*, a biologist would call it *Ursus americanus.*

Common names such as bear, giraffe, canary, and alligator are fine for familiar animals, but many of the animals discussed in this book are not so familiar. The only name many of them have is a name such as *Ursus americanus.* There are many kinds of tapeworms, and just "tapeworm" is not precise enough. Names such as *Ursus americanus* are more specific: The name refers to a particular kind, a *species*, of bear. There are several species of bears. *Ursus americanus*, for example, sets the black bear apart from *Ursus horribilis*, the grizzly bear.

Both the grizzly bear and the black bear are called *Ursus* because they are similar enough to each other to be in the same *genus*, which is *Ursus.* The second word in the name is the species name, such as *americanus.* The polar bear, on the other hand, is called *Thalarctos maritimus.* It is not in the same genus as the other two bears, but shares enough similarities with them to be in the same *family*, which is Ursidae.

These words—family, genus, and species—are part of a hierarchy of classification. Biologists have been classifying things ever since the 1700's, when it was started in earnest by a Swedish biologist named Carl Linnaeus. The science of classification, called *taxonomy*, is much more than a name game. It is an attempt to determine the evolutionary relationships between the living things on this planet.

Living things are placed in groups, ranging from the largest, called *kingdom*, down to the smallest, called *species*. The hierarchy, from largest to smallest group, is kingdom, phylum, class, order, family, genus, species. Any of these can have prefixes such as "sub" or "super" attached to them to express certain relationships. If we trace the bears' classification we find that all bears are in an order called Carnivora that includes all meat-eating mammals. Mammal is the name for animals in the class Mammalia. This class includes people along with all the other animals that, among other things, have hair, bear their young alive (mostly) and produce milk for the young. All mammals are in the subphylum Vertebrata, which includes mammals, birds, reptiles, amphibians, and fish. The kingdom? Animalia, which is just a slightly fancy way of saying animal.

Throughout this book are references to phyla (the plural of phylum), classes, orders, and so on, to which parasitic organisms belong. Particular parasites are referred to by the name that is, in the author's judgment, the most convenient. Quite often the common name of the parasite is the same as the species name. For example, the name of a frequent parasite of man is *Ascaris lumbricoides*. We could just call them roundworms, and they are widely referred to by that name. However, roundworm does not distinguish *Ascaris lumbricoides* from the hundreds of other species of roundworms. Most of the time *Ascaris* is called *Ascaris* in this book, but sometimes they might be called ascarids.

Some of these organisms have fairly long names, such as

Plasmodium falciparum, one of the organisms that cause malaria in people. There are four species of *Plasmodium* that cause malaria. It would get tiresome to say *Plasmodium falciparum*, *Plasmodium malariae*, or *Plasmodium vivax* every time a reference is made to one of these organisms. Therefore, once the name of the genus, *Plasmodium*, has been established, the different species are referred to as *P. vivax*, *P. falciparum*, and so on.

When only one species of a genus is of interest the organism will be known by its generic name only—such as *Ascaris.* If the generic name is a bit unwieldy, such as *Dibothriocephalus latus*, *D. latus* will be employed for brevity's sake, and there are times when a common name such as pinworm will do quite nicely. When referring to several different species, the genus name plus "spp." is used, for example, *Ursus* spp.

Another word problem is when to use *infection* and when to use *infestation.* An invasion by organisms such as bacteria is generally thought of as an infection, and a mess of fleas on a dog is more likely to be called an infestation. One definition is that an infection occurs when the organisms reproduce inside the host's body rather than outside. However, the distinction between outside and inside reproduction is often hazy. Generally the word infection is used in this book to refer to invasions by organisms such as worms. Some people prefer infestation for worms, but infection is coming into greater use.

3.

A Few Constant Companions

Demodex folliculorum

Everybody looks at his or her face in a mirror at least once a day. You might or might not like what you see reflected in the mirror, but certainly you would be, at the least, fascinated and, at the most, horrified if you could see what is probably living there.

There is at least a 50 percent chance that a tiny creature with the impressive name of *Demodex folliculorum*, the hair follicle mite, is comfortably established in your skin. The name tells you where to find *Demodex*. Each hair (and there is hair on everyone's face) grows from a root. The hair root and the material that covers it make up the hair follicle. *Demodex* lives down in the hair follicle. They are also found in the sebaceous glands that secrete skin oils. They are most likely to be found in the skin around the nose and eyes, including the eyebrows. They have also been found in earwax.

Mites are related to spiders and other eight-legged creatures, such as ticks. Most mites look a little bit like spiders, but not *Demodex*. *Demodex* looks more like a worm than a spider. It is about 0.4 millimeters long, has stubby little legs, and a disproportionately long abdomen compared to the rest of its body. It is the long abdomen that makes it look like a worm. They are too small to be seen with the naked eye, but they can be seen with a low-powered microscope. Their presence can be revealed by examining skin scrapings under a microscope, and they are usually covered with yellowish skin oils. If still alive when looked at under the microscope, they will move in a wormlike way, although they are not worms. The oily human secretions covering them and the wormy movement is enough to get a good "yuck" out of most people who see them for the first time, especially if the observer knows that the mites came from his or her own skin.

This is the way Demodex folliculorum *appears in a microscopic view of a thinly cut section of human skin. The long abdomen is easily seen in the one to the right.* YALE UNIVERSITY SCHOOL OF MEDICINE, DEPARTMENT OF PATHOLOGY

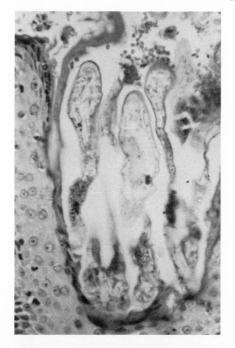

Just what *Demodex* is doing down there in the hair follicle is not at all clear, nor is very much known about its life cycle. Like any other animal, it has to eat something. Its food is probably material in the hair follicle, such as dead cells or oil from the sebaceous glands. It may also eat a little living tissue, but, even if it does, most people who harbor these mites never (or hardly ever) suffer ill effects from their guests' activities.

Since *Demodex* seldom causes symptoms, it is frequently referred to as a "harmless parasite." When harmless, it seems to be more like a commensal than a parasite. Its case for commensal status is weakened by the fact that, in some people, it can cause various symptoms, albeit infrequently. *Demodex* has been implicated in the formation of blackheads and acnelike conditions. These conditions often occur in people who use various cleansing creams instead of soap and water. Rarely, *Demodex* causes a condition is which the skin becomes red, scaly, and wrinkled. Hair is lost, and there may be secondary infections by bacteria.

No one really knows why *Demodex* is on good behavior in almost everybody and wreaks havoc in a few. Many parasitologists believe that *Demodex* produces symptoms only when there are nutritional deficiencies or general poor health. Adding further to the confusion is the knowledge that *Demodex* is more likely to be found in clean, undiseased skin than in dirty, sick skin.

In dogs, *Demodex* can cause a serious, often fatal, infection. It is not known if the *Demodex* that infects dogs is the same species as the one that lives in humans. Some parasitologists believe that dog and human *Demodex* are the same species, while others say they are different. No one has been able to show that dogs can get *Demodex* from people and vice versa. Other animals known to provide a home for *Demodex* are cats, horses, pigs, and cows. In pigs, *Demodex* causes the formation of white pustules.

The life cycle of *Demodex* is a paradox. It is simple but confusing. There is no intermediate host. Females lay eggs that hatch into six-legged larvae which slowly develop into adults. The newly hatched young generally take up residence on the host in which they were born. It is possible that any one person can have a more or less permanent infection made up of what amounts to a *Demodex* family dynasty. It is obvious, however, that the mites spread from person to person. It is known that a few adventuresome *Demodex* wander around on the skin surface. These adventurers may be the ones that strike out to settle new territories. The mite is probably spread by close contact or through items such as towels used by more than one person. There is, however, no pattern to this transmission. A person can live for years in a household full of *Demodex* hosts and never get the infection himself. It is known that some *Demodex*-free dogs can associate freely with infected dogs and not pick up the infection, but others are infected quickly.

Demodex is an example of a fairly well-adapted symbiont. Most of the time, it reacts favorably with the host, and it can spread without the use of intermediate hosts. Compared with most other mites and their relatives, the insects, the rate of reproduction is slow; yet its numbers and range are respectable. Most parasitic organisms produce eggs by the hundreds of millions, and many have evolved mechanisms to get the maximum number of new individuals out of each egg. This huge production of eggs is an adaptation to overcome the unbelievably small odds that any one egg will ever survive to adulthood. The fact that *Demodex*, with its slow rate of reproduction, continues to thrive is evidence of the efficiency of this animal's adaptations.

Escherichia coli

This book is about animal parasites. Although *Escherichia coli*, often simplified to *E. coli*, are bacteria, not animals, they are qualified to be exceptions. They are certainly no exception in your large intestine. The large intestines of all human beings are teeming with this tiny organism.

E. coli is not regarded as a *pathogenic* bacterium—that is, one that causes disease. However, under certain conditions, it can act very much like a pathogen. Most of the time, the *E. coli* and related bacteria in your intestine can be considered to be commensals. They obviously get what they need there, or they wouldn't be there. Some biologists think the real relationship might be a shade mutualistic, but many experiments involving the raising of laboratory animals free of intestinal bacteria have yielded little evidence to support this idea.

If these bacteria get into the urinary tract (the tubes that carry urine from the bladder and kidneys to the outside), they can cause an infection. They can cause nasty infections in the gallbladder, appendix, and sometimes right in their own backyard, the large intestine, particularly the lower part, called the rectum. They can also cause infections in the bloodstream.

Just why *E. coli* lives most of the time as a polite commensal, only to cause disease at other times, is not known among bacteriologists. Infections in the urinary tract are not too hard to understand; the organism is out of its normal environment there. The chemical relationship between the bacteria and the host is different in the urinary tract than in the intestine, and the bacteria behave differently. Not so easy to understand is why they should sometimes act like pathogens in their usual abode. One possibility is that the bacteria that cause the infections are mutants—slightly different from those that are companionate commensals. Conditions in the large intestine can change from time to time, a circumstance that could encourage pathogenic behavior.

Many uses have been found for *E. coli.* It is used widely in laboratories as an experimental organism. It is particularly useful in scientific studies in which large numbers of organisms are needed.

Since *E. coli* are to be found wherever people are found, the bacteria can be used an an index of people-caused pollution. When water is checked for purity, the major test is one for the presence of what are called coliform bacteria. Coliforms are *E. coli* and other bacteria that commonly live in the human gut. The presence of certain levels of *E. coli* in water is an indication of contamination by sewage.

Entamoeba gingivalis

Amoebas are one-celled organisms. Almost everybody who has had a high school or college biology course has seen these fascinating creatures creeping along a microscope slide, changing their shape as they go. There are many different kinds of free-living and parasitic amoebas. Some of the parasitic forms cause serious, often fatal, diseases in humans and other animals.

The amoebas you may have seen in biology class were chosen for visibility. These forms have many grayish granules in them. The granules make them easy to see. The substance of which the amoebas and other cells are made is called *cytoplasm*. Amoebas move by extending their cytoplasm in fingerlike projections called *pseudopods* (Latin for "false feet"). The grayish granules seem to pour into the pseudopods as the amoeba moves. Amoebas also eat with pseudopods. Food is enveloped by the pseudopods and incorporated into the amoeba.

Entamoeba gingivalis lives in human mouths. As the name implies, it is likely to be found in the gums, particularly in gum pockets around the base of the teeth. Unlike the amoebas in biology class, this amoeba is not so easy to see. It is crystal clear, almost glasslike. However, with the proper staining or use of the right kind of microscope (phase contrast microscope), the animal can be seen.

E. gingivalis is known to eat white blood cells, bits and pieces of gum tissue cells, a few red blood cells, and a few bacteria. The amoeba's fare indicates that it is, indeed, a parasite and that it is making its living by eating minuscule bits of you. This amoeba's status, however, is far from clear.

It is found in practically every human mouth at one time or another. It is much more likely to be found in mouths afflicted with diseases such as pyorrhea (a disease of the gums) than in healthy mouths. However, no one has ever been able to prove that *E. gingivalis* causes gum disease or any other disease in the mouth or elsewhere (it is sometimes found in the bronchial tubes leading to the lungs). It is quite possible, indeed likely, that a few of these creatures can be found in the healthiest and cleanest of mouths. *E. gingivalis* is found in 95 percent of diseased mouths and in about ten percent of clean, healthy mouths.

If *E. gingivalis* does not cause gum disease, why is it found in such profusion in diseased mouths? One possible explanation is that there is more for it to eat in a diseased mouth. A diseased mouth can provide a bonanza of food for this tiny creature: bacteria, of course, and white blood cells that have come along to eat the bacteria.

So *E. gingivalis* remains an enigma. Is it a harmless commensal or a harmful parasite? Or is it usually a commensal sometimes turned disease-causing parasite? There is a possibility that the white-blood-cell-eating habit may be a minor problem. White blood cells help to fight off infection by ingesting bacteria. If *E. gingivalis* eats the white blood cells, there will be fewer of them around to dispatch bacteria. On the other hand, *E. gingivalis* might be of some benefit by helping to clean up the "debris of battle"—the remains of white blood cells and bacteria at infection sites.

Not too much is known about the life cycle. There is no intermediate host, and the amoebas do not go into inactive *cyst* stages as is the case with other parasitic amoebas. It is fairly certain that the amoeba is spread by kissing, particularly the kind of kissing called "soul-kissing" or "French-kissing," in which the kissers roll their tongues inside each other's mouths.

If you are worried about *E. gingivalis* creeping around in your mouth, and there seems to be very little reason for you to spend too much of your worrying time on them, you can control their numbers by keeping your mouth clean. No extraordinary measures are needed. The usual practices of oral hygiene, such as flossing and brushing your teeth (particularly after meals) and the occasional use of a mouthwash are more than sufficient. You don't have to spend a lot of money on name-brand mouthwashes, either. Rinsing out with a weak saltwater solution from time to time will accomplish the same results as rinsing with a mouthwash.

Demodex, *Escherichia*, and *Entamoeba gingivalis* are reminders that the divisions between commensalim, mutualism, and parasitism are not always clear-cut. They are also examples of situations in which the symbionts—host and guest—are usually well adapted to each other.

4.

Roundworms

To most people a worm is any animal, more or less cylindrically shaped, that moves in a wormlike fashion: squirming, wiggling, oozing, inching, and so on. Worms produce reactions of revulsion and disgust in many people. Perhaps worms remind us of snakes, and snakes, depicted as the symbols of evil in the Book of Genesis, have never been particularly well-liked by mankind. Wormy things are thought of as being slimy. Many worms, notably the familiar earthworm, are indeed slimy, but snakes are definitely not slimy. Worms are also associated with death. The "worms" often seen in decaying dead bodies are not worms at all, they are maggots—fly larvae. Contrary to a fairly widespread myth, earthworms do not invade coffins to eat the flesh of the deceased.

Not every animal that is more or less cylindrical in shape and moves in a wormy way is a worm. Many animals dismissed as worms by the casual observer are not really worms at all. They might be caterpillars or maggots, which are immature forms of insects. To a biologist, a caterpillar is no more a worm than you are a starfish. However, not too long ago even biologists made little distinction between worms. About a hundred years ago anything that looked like a worm was likely to be classified as belonging to a phylum called "Wurmen," a word derived from the German that means, appropriately enough, "worms" and from which we get the word "vermin."

As biologists examined worms more closely, they found that there were vast differences between such creatures as a parasitic roundworm and an earthworm. An earthworm is really a far more advanced animal than a roundworm or a flatworm. In many respects human beings have much more in common with frogs than earthworms have with round-worms and flatworms. Earthworms, by the way, are "re-spectable" free-living citizens. Contrary to some of the folk-lore you may have heard, earthworms are never parasites. They live in the soil, burrowing through it, obtaining food as they move. As they move, they churn up the soil, loosening it and exposing it to air. They are regarded as useful to mankind since their activities tend to increase the fertility of the soil. They are also well thought of by fishermen as the ideal bait. They have been proposed as an excellent source of protein—considerably cheaper than beef or lobster. If you are interested, there are earthworm recipe books available.

There are three major phyla of "true" worms or *helminths*, as they are often referred to by biologists, particularly parasitologists. There are also other, smaller worm phyla. The major worm phyla are Platyhelminthes, the flatworms; Nemathelminthes, the roundworms; and Annelida, the seg-mented worms. "Platy," by the way, means flat, and "nemat" means threadlike. The platyhelminths and nema-thelminths include many parasitic species. The only annelids that are of any importance as parasites of humans or domestic animals are leeches. Another group of worms that includes parasites of domestic animals and occasionally humans is the Acanthocephala, the spiny-headed worms.

Not all biologists agree with the classification briefly outlined above. Many biologists prefer to consider the roundworms as a class within a larger phylum called Aschelminthes. Within this phylum the roundworms are the class Nemathelminthes or Nematoda. The latter is often preferred if for no other reason than that it takes less time to say. Also included in this phylum are the class Nematomorpha (horsehair worms) and the Rotifera. The latter are animals called rotifers, common inhabitants of water, both fresh and salt. Microscopic examination of practically any sample of pond water will usually reveal a few of these interesting animals. None is known to be parasitic to humans or to any animal of economic interest to people.

Many nematodes are parasitic to people and to economically important plants and animals. No less than fifty species can parasitize people, and about a dozen or so of these are frequent inhabitants of one part or another of the human body. Quite a few species, probably about half of all those that exist, are free-living. Even some of the free-living forms can be troublesome, fouling water supplies and eating the roots of crop plants and lawn grass. Brown patches on lawns are quite often due to nematodes.

Roundworms, both parasitic and free-living, are found just about everywhere. If you stand on the ground, there could well be a few hundred thousand of them in the soil directly under your feet, and there might be a few, or a hundred or so, living inside you. Parasitic forms tend to be larger than free-living forms. Most free-living nematodes are never seen, except by those who are interested in them. They are micro-scopic in size, although not so small that a low-powered microscope can't make them visible. About one half milli-meter in length and one twentieth of a millimeter in thickness is a typical size for nematodes that live freely in soil, water, beer kegs, garbage dumps, and just about any place imagin-able. Parasitic roundworms are more variable in size, ranging from the microscopic to monsters that are scores of inches long.

Most nematodes, of whatever size, look pretty much like any other. They all have cylindrical bodies, tapering to points at each end. Their bodies are covered with a tough, protec-tive coating called the *cuticle*. Practically all are white, al-though some have a grayish, pinkish, or yellowish cast to them, and at least one species is a bright red, a circumstance made even more startling by the fact that this red one is one of the largest known parasitic roundworms.

Roundworms have muscles running the long way on their bodies. This arrangement allows them to move back and forth in a characteristic whiplike way. Once you have seen a nematode whipping its way across the field of view of a microscope, or writhing in a mass just expelled from a sec-tion of pig's intestine, you will always be able to identify a worm as a nematode by the way it moves.

Although most nematodes are white, tapering cylinders, some have a few modifications of the basic cylinder. Many of these modifications are adaptations for parasitic living. Hookworms, for example, have an impressive array of "teeth" at the mouth end. They use these "teeth" to latch onto the intestine of the host.

Structurally, nematodes are a bit more sophisticated than

flatworms. Nematodes have a complete digestive tube: There is a mouth at one end of the tube, through which food enters, and an anus at the other end, through which wastes pass to the outside. Flatworms have only one opening in a pouchlike digestive cavity, and some don't even have that.

In practically all species of nematodes there are separate male and female worms. This statement may sound odd, but there are many animals, particularly those generally called worms, in which the male and female reproductive organs are in one individual. Such animals are called *hermaphrodites*. Earthworms are hermaphrodites, as are most flatworms. The reproductive apparatus of a female roundworm is formidable, capable of producing eggs by the hundreds of thousands. Massive, prodigious rates of reproduction are the rule for parasitic nematodes. Very few of those eggs, however, make it to a new host.

No one knows exactly how many animals, including people, harbor roundworm parasites. In 1947, a famous parasitologist named Stoll in an article called "This Wormy World" put forth the estimate that two billion human beings are infected by roundworms. There are more people now than there were in 1947, so that number is probably greater today, although the percentage of the world's population infected with roundworms is probably less than it was in 1947. In all probability, approximately half the world's human population has roundworms at one time or another.

Of course, humans are by no means the only host to roundworms. Practically every pig in the world harbors them, and almost every dog is born with them well established in its intestinal tract. There are few species of vertebrate animals (animals with backbones—fish, amphibians, reptiles, birds, mammals) that are not natural hosts to several species of nematodes. Nematodes are truly one of the most pervasive, but inconspicuous, forms of life on this planet.

As parasites they are opportunists, establishing in practically any part of the body. There are species of nematodes that live in intestines, blood, muscle, skin, mouth, eyes, brain, lungs, heart, stomach, kidneys, liver, and in just about every cavity, nook, and corner that can afford some degree of shelter and sustenance.

In this chapter we will investigate the private lives of some of the more common roundworm inhabitants of humankind.

Ascaris lumbricoides

When I was teaching anatomy and physiology to students of X-ray technology, I achieved what amounted to a minor diagnostic triumph. Every week in the hospital where the students worked there was a "diagnosis of the week." A particularly interesting X ray was pinned to a lighted X-ray reading panel and passersby were invited to offer a diagnosis. The triumph occurred when the X ray of the week was a barium X ray. The patient had swallowed a thick milkshake kind of beverage that contains a barium compound, a substance that will show up on the X ray. The procedure makes the digestive tract—stomach, small and large intestine—visible on the X ray.

In this particular barium X ray there was something that most definitely did not belong there. Clearly visible in the small intestine was an object shaped like a long, bent stick, curved at one end in a shape somewhat like a shepherd's crook. When I first saw the picture I was with a group of doctors and instructors. No one in this group, including the doctors, could identify the sticklike thing in the patient's intestine.

"That's an *Ascaris* worm," I said.

"What's an *Ascaris?*" was the immediate rejoinder from several members of the group, a not surprising response. Not too many people in the United States have ever heard of this organism.

"Oh, yes," or words to that effect was the response from some of the doctors in the group, who had not immediately associated the "stick" with the word *Ascaris*, which they may have heard once or twice in the course of their medical education.

Of course, I was not the only person in the hospital who knew what an *Ascaris* was. The patient's doctor, who had ordered the X ray, knew that the "stick" was an *Ascaris* and assumed, correctly, that an *Ascaris* was sufficiently unfamiliar to many American doctors to make the picture a particularly interesting diagnosis of the week.

So what *is* an *Ascaris*? *Ascaris*, or ascarids, as they are also referred to, have been in the company of man for a long time. No one knows quite how long, but *Ascaris* or some similar worm probably found shelter and sustenance in the intestines of our prehuman ancestors.

Ascaris was probably the first internal human parasite people knew about. It is fairly large as worms go. The adult worms range in length from about eight inches to twelve inches, and are about as thick as a pencil. When they sicken or die, they are sometimes passed out in the stool. So large an organism in the midst of material that came from within one's own body could not have escaped the notice of our ancestors, however primitive.

Ascaris possibly gained its introduction to the human interior when humans started to domesticate pigs. Pigs are commonly infected with ascarids that are indistinguishable from the human variety, but, however alike they may be in appearance, pig *Ascaris* cannot live in humans and human *Ascaris* does not get established in pigs. Ten thousand years or so ago, when people first started to keep pigs, some of the eggs of the pig ascarids undoubtedly found their way into humans. Through the process of natural selection, some ascarids became adapted to live in the human intestine. That evolution continued until the two separate varieties of *Ascaris* developed. Not all parasitologists agree with this hypothetical evolutionary scenario, maintaining that humans had their own ascarids long before their cultural evolution carried them to the point where they domesticated animals. Although pig ascarids usually cannot become established in people, they can still be a health problem under certain conditions.

For years *Ascaris* was not regarded as a health problem of any particular consequence. That attitude was not exactly correct. They may or may not be a health problem depending on a number of factors, including how many worms a person has, and the general condition of the host. About a hundred years ago, it was widely believed that these white-to-pinkish worms, so often seen in the bowel movements of children, were the child's guardian angels. While these worms usually do little to draw attention to their presence, and some people can harbor them for years and remain in robust good health, they are capable of wreaking some awful havoc. "Guardian angel" seems hardly appropriate in reference to ascarids.

LIFE CYCLE

The adult worms live in the small intestine, where they feed on semidigested food. The females constantly produce eggs. A single female ascarid can produce as many as 200,000 eggs a day. The average egg-laying life of a female *Ascaris* is about ten months. A little arithmetic will show that the potential egg production of a female ascarid is about 60 million eggs. At any one time the body of a female ascarid may contain as many as 27 million eggs.

The eggs pass out of the host in the feces (intestinal wastes). In areas of the world where there are no toilets or where people have careless habits, the eggs are likely to find their way to the ground. The eggs have a thick, crusty shell that protects them from many of the hazards in the great world outside the host. They do, however, require a bit of moisture and reasonably warm temperatures if they are to develop to the stage where they can reinfect another host. Extreme dryness will kill them, and cold will retard their development but not necessarily kill them. Eggs have stayed alive in refrigerators four years and longer.

The eggs, after being swallowed by a new host, hatch into tiny (14 micrometers; a micrometer—formerly called micron —is 0.001 millimeter) larval worms when they get to the small intestine. At this point it would seem that the larvae would be busy getting established, growing into adults, and producing more eggs. However, rather than settle down immediately, the larval worms burrow into capillaries in the wall of the intestine, from where they are carried into the general blood circulation.

Eventually they are carried to the lungs, where they become lodged in the alveoli, the tiny air sacs in the lungs. In the air sacs they grow to lengths of one to two millimeters. When they reach this size they break out of the air sacs and move into the bronchial tubes and on up into the trachea (windpipe) and pharynx, the area in the back of the mouth where the windpipe, esophagus (food pipe), and nasal passages meet. In the pharynx they are reswallowed (at least most of them are) and returned to the small intestine. This time, their need to travel apparently satisfied, they settle down, grow to adult size, and start to produce eggs. About three months is the usual time between the ingestion of eggs and the start of egg production by the adult worm.

The *Ascaris* life cycle is fairly simple and straightforward. There are no intermediate hosts; reinfection is through the most direct route, the mouth.

SYMPTOMS

Symptoms are widely variable. Most of the time, *Ascaris* behaves as a well-adapted parasite and causes the host no discomfort. Just how much discomfort, if any, the host will suffer is dependent on factors such as the number of worms, the body's allergic reaction to the waste products of the worms, and the amount of physical damage the worms might do. Of course, all of these are directly related to how many worms there are. One worm probably won't produce enough waste products to bother the host, but a hundred worms just might. Also, the worms don't live forever. Twelve to eighteen months is the usual lifespan. When they die they decay, and the decay products can cause allergic reactions in some people.

Large numbers of larval worms moving through the lungs can produce symptoms similar to pneumonia. Occasionally, the larvae wander through organs other than the lungs, producing a variety of symptoms, depending on which organs are affected. They can get into the brain and other parts of the nervous system, causing symptoms ranging from dizziness to epilepticlike seizures and paralysis of varying degree.

As long as the worms stay in the small intestine and eat nothing but semidigested food, and there aren't too many of them, they will hardly be noticed. Occasionally the host might feel some vague discomforts, such as a mild bellyache. Some parasitologists believe that the worms vary their fare of semidigested food by taking little nips at the wall of the intestine. They do have the right mouth structure to take a little nibble now and then. Such activity may or may not cause discomfort or loss of blood, depending on how often it occurs.

Occasionally the worms will do things ranging from the annoying to the bizarre to the extremely dangerous. Worms agitated by lack of food or irritated by something the host has taken in, such as drugs, can in their agitation move into a mass and actually block the intestine. Often surgery is required to clear up the blockage, but there are medications that will usually clear the worms out.

Sometimes a wanderlust comes over the worms and they start to "explore." They can wander through the bile ducts and into the liver. In so doing they can block the bile ducts, causing pain similar to that of gallbladder attacks. Every now and then an ascarid wanders up into the stomach, esophagus, and on up into the mouth, where it may emerge from the mouth or nose of an understandably terrified host. While an experience of this kind is enough to bring on hysterical behavior in the most stouthearted person, it is harmless.

Dead or dying ascarids emerging from the other end of the body during bowel movements can also throw the surprised host into a panic. Actually, passing the worm in this manner has its beneficial aspects. It lets the host know of his or her unwanted guests and the dead worm won't decay inside the host and produce possibly harmful breakdown products.

Although worms emerging from the mouth or anus are not dangerous, those that pierce the intestine and crawl into the body cavity are another matter. Piercing of the intestine

Ascaris worms don't always stay in the small intestine. This one was found in a human liver removed in an autopsy. The worm, which apparently caused a severe abcess in the liver, can be seen hanging below the liver specimen and projecting into it.

THE CENTER FOR DISEASE CONTROL

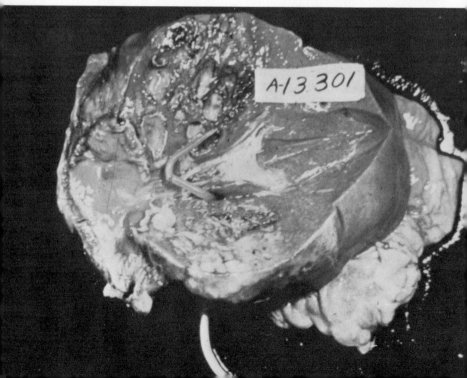

by the worms can, and usually does, cause peritonitis, an infection of the inner lining of the body wall, called the peritoneum. Peritonitis can be fatal, but it can usually be treated successfully with antibiotics and other measures. Ever so rarely, a worm that has been wandering in the body cavity might emerge through the umbilicus (belly button). These occurrences, however, are rare.

TREATMENT

Treatment of *Ascaris* infection is a fairly simple matter. A drug called pyrantel pamoate (Antiminth) is one of the most effective treatments available. All the patient does is swallow the medicine; no special preparations are needed. If the worms have caused a blockage of the intestine, medications are sometimes given via a tube passed in through the nose, pharynx, and esophagus down into the intestine. There are many effective drugs available and the choice is largely a matter of what else might be there. Some species of worms are killed by one drug and merely irritated by another. So a complete diagnosis is usually advisable before treatment is given. In some parts of the world, particularly tropical areas, infection from more than one kind of worm is the rule rather than the exception. Diagnosis of intestinal worm infections is made by checking the feces for eggs.

FACTS AND FIGURES

Ascaris is found just about everywhere, but it is found more in some places than in others. Parasitologists estimate that 900 million to one billion people in the world have *Ascaris* at any one time. About a million of these are in the United States. About two percent of the total population of

North America harbors *Ascaris.* In Asia (other than the U.S.S.R.) the figure is 35 percent; in Africa and tropical America some 30 percent of the population has *Ascaris.*

Ascaris seems to seek out the poor and the under-privileged. Of course, the worm really doesn't do any seeking at all, but it thrives in the conditions in which most poor people throughout the world are forced to live. It also seems to have a propensity for children between the ages of two and nine, but this propensity is due to the habits of children rather than to any preference of the worm for them.

In the crowded slums and shantytowns that exist on the fringes of cities in Africa, Asia, and tropical America, sanitation and plumbing are notably nonexistent. In these conditions children, in particular, are likely to relieve themselves on the ground, the same ground on which they play. The absence of pavement and the presence of a warm, wet climate are favorable for the development of the eggs. Children, playing in the dirt, carry the eggs to their mouths on dirty hands. Adults are by no means safe from infection. The massive concentration of worm eggs in such situations makes the infection of everybody almost inevitable.

In many parts of the world "night soil," a euphemism for human wastes, is used for fertilizer. Economically this practice makes sense. Commercial fertilizer is expensive and many poor farmers do not have enough animals to produce a sufficient amount of manure from that source. However, the use of human wastes for fertilizer spreads many kinds of parasites, not just *Ascaris.* In the case of *Ascaris*, the eggs can be ingested in vegetables and other plant food that has been grown with human wastes.

The prevalence of worms such as *Ascaris* among the poor

sets up a vicious cycle. The worms rob the people of the little nutrition that is available to them. Robbed of food from within, the people have less energy to do the work that might help them to get out of a life of poverty. Malnutrition aggravates the effects of *Ascaris* and encourages the worms to wander through the body, often producing the disastrous results discussed earlier. A massive amount of money spent on sewage systems, medications, and education is needed to control *Ascaris* in those areas where its incidence approaches 100 percent.

In the United States, *Ascaris* has been concentrated in mountainous, agricultural regions of the South, particularly in the region called Appalachia. Again, the reason is lack of modern plumbing and a tendency of children to use the ground rather than walk up the hill to the outhouse. Increased use of plumbing has greatly reduced the incidence of *Ascaris* in the United States. However, success often brings the danger of complacency and carelessness. The worm can and has made comebacks. When I taught high school biology I stressed that point as I lifted a fourteen-inch ascarid out of the bottle of preserving fluid with the comment, "This is why your mother always told you to wash your hands before eating."

Worldwide, the future of *Ascaris* and other intestinal parasites is largely a matter of economics. In many parts of the world where *Ascaris* is prevalent, such as Africa and tropical America, the human population continues to increase at an alarming rate. People from the countryside (many of whom already have *Ascaris*) are drawn to cities, where they concentrate in festering shantytowns. Such conditions are made to order for *Ascaris*. If nothing is done about relieving these conditions, *Ascaris* and similar parasites will continue to flourish, as they always have, in the backwash of human civilization.

Trichuris/trichiura

This worm thrives in the same conditions that encourage the proliferation of *Ascaris.* Dual *Ascaris-Trichuris* infections are quite common in areas and situations where *Ascaris* is prevalent. *Trichuris* is considerably smaller than *Ascaris.* The females are about an inch and a half to two inches long. They are thin enough to merit the description "threadlike." They are called whipworms because they look like buggy whips. The posterior or rear portion of the worm is thick and is the "handle" portion of the whip. The anterior portion of the worm is thin and is the "whip" portion.

LIFE CYCLE

Eggs of *Trichuris* are taken in pretty much the same way as those of *Ascaris*—on dirty hands contaminated by egg-infested soil. When the eggs hatch in the small intestine, the young worms get right down to business. Unlike *Ascaris,* they do not go on any trips, but stay in the intestine. They move downward in the digestive tract until they reach a part of the large intestine called the caecum. This is the pouch of large intestine just below the point where the small intestine joins the large intestine. The appendix branches off from the caecum.

The worms weave the thin, anterior "whip" portions into the mucosa (covering) of the caecum. Just exactly how they feed is not too well known. However, it is known that whatever way they obtain nourishment does the host no good. It is thought that the worm secretes enzymes that chemically break down the cells with which it is in contact, and that it eats the products of the cellular disintegration. The females lay from 3,000 to 10,000 eggs a day. These eggs pass out in the feces and continue the life cycle.

SYMPTOMS

Although *Trichuris* is considerably smaller than *Ascaris,* it can cause much more damage. As is the case with *Ascaris,* children are frequent victims. A few worms will not cause noticeable symptoms, but heavy infections, when the number of worms exceeds two or three hundred and, in some instances, three or four thousand, can be quite serious and even fatal. Symptoms of heavy infection include anemia, abdominal pain, nausea, blood in the stool, headache, and fever.

TREATMENT

Diagnosis is made by finding the eggs in the stool. The eggs have an oval shape that is particularly easy to recognize. To many observers the eggs look like tiny lemons.

Treatment is similar to that for *Ascaris.* Medication is taken by mouth. In the past the medication was introduced into the large intestine through a tube inserted into the rectum.

FACTS AND FIGURES

Estimates of the number of people with *Trichuris* hover around 500 million. Infections in the United States probably number around 500,000 and are concentrated in the same areas as *Ascaris.*

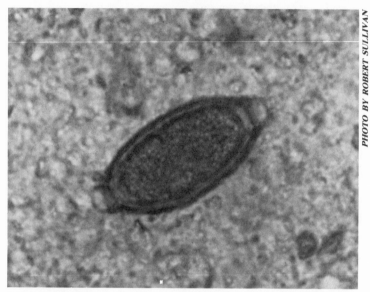

PHOTO BY ROBERT SULLIVAN

A Trichuris *egg found in the feces of a monkey. Human* Trichuris *eggs are practically indistinguishable from this one: The egg might even have been produced by the same species that infects man.*

YALE UNIVERSITY SCHOOL OF MEDICINE,
DEPARTMENT OF COMPARATIVE MEDICINE.

Enterobius vermicularis

A continuing argument among parasitologists is just which human parasite has the dubious honor of infecting the most people. Although *Ascaris* probably takes the overall prize, *Enterobius* is most likely the champion in the United States today. The only worm that might ever give it any competition for top spot in the U.S. and Canada is *Trichinella* (page 86).

Estimates vary, but the educated guess for the United States and Canada is eighteen million at any given time. The number of infections worldwide is around 210 million. However, many parasitologists insist that practically *everybody* in industrialized nations in temperate parts of the world has had *Enterobius* at one time or another. Fortunately this worm is, compared to most others, rather innocuous. Few people who harbor this organism are ever aware of it. Many parasitologists think it should be regarded as a commensal rather than a parasite.

White to yellowish-white in color, they are quite small, ranging in size from eight millimeters to thirteen millimeters long (one-third to one-half inch) and about four tenths of a millimeter in diameter. The posterior of the female is clear and pointed; hence the common name "pinworm."

LIFE CYCLE

The life cycle of *Enterobius* is simple and direct. The adults live in the large intestine, including the caecum. They might also establish in the lower part of the small intestine near the caecum. When the female is filled with eggs (about 11,000), she moves down to the rectum and comes out at the anus. She lays her eggs on the skin in the region around the anus, the perianal region. The worms dry out in the process and sometimes literally explode, scattering eggs as they do so. The eggs enter a new host through the mouth, usually. Sometimes the eggs hatch on the skin in the perianal region and the newly hatched worms crawl back into the large intestine through the anus.

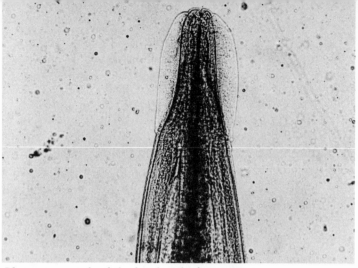

Photomicrograph of the head end of a pinworm

THE CENTER FOR DISEASE CONTROL

SYMPTOMS

The egg-laying activities of the worm often produce an intense itching in the perianal region. However, itching is a frequent occurrence in that area, and few people ever associate so common a sensation with the possibility that they might have worms. The itching might cause loss of sleep which in turn can bring on irritability, restlessness, and a general listlessness. In children the worm can bring on squirming, fidgeting, and nose-picking.

TREATMENT

With *Enterobius* a discussion of treatment seems hardly worthwhile. Ordinarily the worms die and disappear in six to eight weeks, but reinfection almost always occurs once they get started. There are many drugs that will kill them quite readily, but the drugs do not prevent reinfection, and reinfection does indeed occur.

Pinworm is one of the most contagious worms known to exist. It spreads readily from person to person, and a host generally gives it back to himself. Itching, anywhere, usually produces scratching. While social custom generally discourages scratching of the perianal area in public, such niceties will seldom prevent a young child from relieving an itch no matter where the itch, or the child, might happen to be. The scratching tends to get eggs on the hands, a circumstance that facilitates self-reinfection. The scratching also tends to scatter the eggs into the air, and that's when the fun begins.

An aware person can prevent such misfortunes as *Ascaris* infection through personal cleanliness. Washing hands thoroughly with soap and hot water is an extremely effective way to prevent *Ascaris* and many other diseases. Not so with pinworm. You can get pinworm by simply breathing.

Typically, pinworm is brought into a house by young children. Children usually pick up pinworm in school. The squirming, scratching, and fidgeting of infected children tends to spread the eggs into the air, and the concentration of eggs in a close situation such as a classroom can be prodigious. Nursery schools and kindergartens are particularly rich sources of pinworm eggs.

The child brings the eggs home on his person, and soon everybody in the family is infected. Once the worms get established in a household only a lot of hard work can get rid of them, and there is no guarantee that anything will break the cycle of infection. The eggs get into everything— into clothes, bedsheets, and blankets, into dust that settles on everything and anything. Eggs are deposited in bathtubs when infected individuals take baths. Everyone else who uses the bathtub can then become infected. The common housekeeping practice of shaking out bedsheets and blankets is a very effective way to spread the eggs all over the place. The general level of housekeeping has little to do with the chances of their spread through the family. They can be as much at home in a spotlessly clean house as in one where the housekeeping is of a more casual nature.

A popular diagnostic method is the so-called "Scotch-tape test." A piece of transparent tape is looped onto a tongue depressor, sticky side out. The sticky side of the tape is pressed to the perianal region of the suspected host several times. The tape is then pressed onto a microscope slide and examined for the presence of eggs. Quite often tests such as these are unnecessary. The worms are often seen on the perianal skin or in the underwear of the host.

Once the presence of pinworms in a household has been established, the first thing to do, if you are at all concerned about getting rid of them, is to stop taking baths and take showers instead. Infected young children should wear tight-fitting pants to help prevent picking up eggs on the hands from scratching. Trim the fingernails of young children and wash their hands often. Everything that comes into contact with infected individuals, such as clothes and bedding, should be washed daily. Handle bedding carefully to avoid scattering the eggs. Protect food from exposure to dust.

The spreading of a salve or ointment on the perianal region will help to prevent the spread of eggs. Be sure to wash your hands *thoroughly* after applying the salve.

As implied earlier, all of these measures might prove to be futile. The infection will probably go away by itself or at least subside after a while in any case, particularly over the summer when reinfection from schoolrooms does not occur, more time is spent outdoors, and the heat will cut down the life of the eggs.

FACTS AND FIGURES

Fortunately for all of us, pinworm eggs are not very hardy. At room temperature (about 20° C. or 70° F.) and a relative humidity of 35 percent, fewer than 10 percent of the eggs survive for more than two days. At summer temperatures and relative humidity of around 40 percent, fewer than 10 percent of the eggs will survive for as long as three hours.

The lethal effect of high temperature on pinworm eggs helps to explain why *Enterobius* is relatively rare in tropical regions, where most other parasites thrive. Pinworm is the parasite of industrialized, temperate nations. It is a democratic parasite, just as likely to be found in the mansions of the rich and powerful as in the hovels of the poor. For many of us it is the only parasite we will ever have, and it would seem that *Enterobius* and humanity are going to continue to be close companions for a long time.

Hookworms

Hookworms could be called the vampires of the worm world. They feed on the blood of their hosts. While no longer a serious health problem in the United States, they continue to be a problem in many parts of the world.

There are many species of hookworms, but only two are frequent human parasites. The two species are the Old World hookworm, *Ancylostoma duodenale*, and the so-called New World hookworm, *Necator americanus.* The name means "American killer."

The designation *"americanus"* is somewhat incongruous.

The worm did not originate in tropical America and the southern United States where it is found today, but it was first discovered in these areas. The worm originated as a human parasite in western and central Africa, where it still exists. It was brought to the New World in the bodies of African slaves. You may have heard of dysentery being referred to as "Montezuma's revenge." *Necator americanus* could well be the African's revenge for the cruel slave trade that flourished until a little more than a century ago.

Once the hookworm became established in the southern United States, a curious evolutionary irony occurred. *Necator americanus* soon became the scourge of the American South, infecting hundreds of thousands of people with devastating results. The victims, however, were mainly white people, particularly poor whites, many of whom harbored deep hatred for the descendants of the black slaves with whom they were in economic competition. Blacks got hookworms, too, but most blacks were resistant. They harbored few worms, and their symptoms were mild, but many of their white neighbors literally withered away from heavy hookworm infections.

The spread of *Necator* from Africa to the western hemisphere by the slave trade is but one example of how the migrations of the human animal have spread parasites from one part of the world to another. *Necator americanus*, for example, has spread not only to America but also to Australia, India, southeast Asia, and China. The Old World hookworm was well established as a human parasite in biblical times and before. It was probably a common parasite of ancient Egyptians. The existence of hookworm could well have been responsible for the writing of biblical passages such as:

And thou shalt have a place also without the camp whither thou shalt go forth abroad. And thou shalt have a paddle among thy weapons; and it shall be when thou sittest down abroad, thou shalt dig therewith, and shalt turn back and cover that which cometh from thee.

Deuteronomy 23:12–13

It would seem that the ancient Hebrews had an idea of how hookworms and other intestinal parasites were spread around. However, hookworm was not known to exist until 1838.

The Old World hookworm is still found primarily in north Africa and southern Europe. But it too has spread with the travels of mankind. It is found today in northern China and in Japan, the west coast of South America, and to a small extent in other areas where *Necator americanus* is the predominant species.

Hookworm is one of the most economically devastating parasites. It attacks humans and many domesticated animals. There are hookworms of cattle, sheep, and goats that severely affect the health of these animals and reduce their value. Dogs and cats are also parasitized by hookworms. While these do not normally infect humans, they can cause an uncomfortable condition when they try (*see* Chapter 9).

LIFE CYCLE

The life cycles of the two human species are quite similar. The worms live in the small intestine. The females of *Necator* are about 12 millimeters long (about one-half inch) and about six-tenths of a millimeter in diameter. *Necator* males are a bit smaller. *Ancylostoma* is a little bigger than *Necator.* The worms latch on to the interior wall of the small intestine with a vicious-looking mouth apparatus. The mouth of *Necator* is equipped with cutting plates, while *Ancylostoma* has mouth equipment that can be more accurately described as teeth.

The worms remain more or less permanently attached. They might, however, change positions from time to time. While latched onto the intestine they suck blood and other body fluids.

Ancylostoma females produce about 30,000 eggs a day, whereas the smaller *Necator* females can manage only a mere 10,000 or so a day. The eggs pass out in the feces of the host.

If the eggs are deposited in warm, moist, loose soil, the larvae that hatch have an outside chance of finding a new host. The tiny (about 0.3 millimeter long and 0.015 millimeter thick) larvae feed on bacteria and soil debris. They grow to a length of a little more than half a millimeter in about five days, if they survive that long. At this time, they stop eating and come to a critical point in their lives. They assume a position in which they extend straight up from the ground, and they wait. They can last about six weeks in ideal conditions. If they don't find a new host within that time they will die. The lucky ones will be stepped on by a barefoot human being or otherwise come into contact with bare human skin. The larvae will then bore into the skin of the prospective new host until they reach tiny veins near the surface of the skin. They are then carried in the blood circulation to the lungs.

In the lungs they break out of tiny blood vessels and get into the air sacs. They move up into the bronchial tubes, trachea, and into the pharynx where they reach another critical point in their lives. They might be swallowed, in which case they will grow into adult worms and continue the species. Or they might be spit out, in which case they will perish.

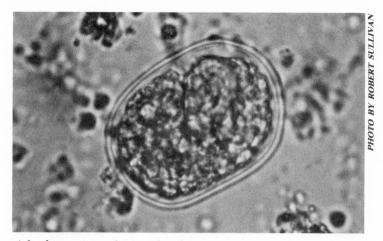

A hookworm egg about to hatch

I spent my childhood in Georgia and can recall the time a playmate offered me some chewing tobacco with the advice that "chewin' tebakky is good fer yer. It'll keep yuh fum gettin' worms." There was a bit of truth in that folksy piece of advice. Tobacco chewers of necessity spit, and the more one spits, the more likely it is that any hookworm, *Ascaris* , or other worm larvae wandering up from the lungs will be ejected into the cruel inhospitable world, rather than descend into the warmth, security, and comfort of one's small intestine. Tobacco chewing was and still is a widespread practice in the South. The tobacco chewing habit has its drawbacks (one soon learned, for example, never to sit by the open window of a moving streetcar behind a tobacco chewer), but it has undoubtedly served to reduce the number of worms that get established in individuals exposed to the larvae. The chewing of betel nut has accomplished much the same thing in Asia.

SYMPTOMS

As is the case with most parasites, a few hookworms will do no great harm, but hookworm infections are all too frequently massive. From the moment they come into contact with the skin these small worms make trouble, ranging from petty annoyances to death and debilitation. When they bore into the skin they cause an itch, long known as "ground itch" or "dew itch" in hookworm country. In the lungs and bronchial tubes the larval worms bring on symptoms similar to pneumonia.

The main trouble from hookworms comes when they are in the intestine, and that trouble comes from loss of blood and the consequent loss of iron. One worm does not take a significant amount of blood. One *Necator* takes about 0.03 milliliter of blood a day. *Ancylostoma* takes more (0.15 to 0.26 milliliter). However, loads of 500 to 1,000 worms are not uncommon. A daily blood loss of from 15 milliliters to 260 milliliters could and usually does result in anemia. (A blood loss of 260 milliliters a day comes out to about a quart a week!) The situation is complicated by the fact that many victims of hookworm infection are poor people who might also be suffering from malnutrition, particularly protein deficiencies, since protein-rich foods generally cost more than other foods. The lack of protein aggravates the effects of the anemia and reduces the natural resistance of the host to additional infection. The substances in the body that fight off infections are various kinds of proteins. A lack of protein means fewer infection-fighting substances.

A high-protein diet replaces many of the vital substances lost to the blood-sucking activities of the worms. Unfortunately, hookworms occur mostly in areas where the diet tends to be low in protein, and where many people don't wear shoes because they can't afford them, don't want to, or because local custom decrees barefoot as the way to go.

Victims of heavy hookworm infection tend to be weak, listless, and lethargic. They are frequently pale and haggard-looking. For years many of the rural people in the southern United States had a reputation as a lazy, shiftless lot. Many of them were indeed shiftless and hardly had the energy to get through a day. The jokes and folklore surrounding the lazy hillbillies of the American South evolved from the very real effects of hookworm infection. These people were frequently shown in movies, comic strips, and other media as a scrawny, lanky, ignorant people interested only in lying around and waiting for the corn mash to be boiled down to whisky.

Heavy hookworm infection can have a particularly devastating effect on children. A child can have just as many hookworms as an adult, if not more (since children are more likely to go barefoot than adults), and a blood and iron loss that does not bother an adult can be burdensome on a small child. Children, particularly those who subsist on a low-protein diet, can suffer stunted growth, delayed puberty, and mental retardation from a heavy load of hookworms. Heavily infected children tend to make little, if any, progress in school, a circumstance that contributed to the "ignorant" label applied to many hookworm victims.

Many victims of hookworm have an interesting, and never fully explained, craving for coarse, gritty substances. These people were, and still are, called "dirt-eaters." However, the craving is not limited to dirt but also includes plaster, clay, and pulverized brick and concrete. This practice is particularly harmful to teeth, which may be ground down to the gum. Another curious effect of hookworm infection is the "reverse iris-dilation reflex." Ordinarily the pupil of the eye will contract if a light is shone on it. In hookworm victims, the pupil gets bigger when light is directed to it. Although the cause of this effect is not known, it is useful in the diagnosis of hookworm infection.

TREATMENT

In 1902, the United States Public Health Service conducted a survey of the hookworm situation in the South. Thousands of such surveys were, and still are, conducted by various government agencies. Most of these reports attract little attention from the general public. Every now and then, however, one becomes big news and such was the case with the hookworm report of 1902. This survey, conducted by an employee of the United States Public Health Service named Stiles, attracted a great deal of attention.

In his report Stiles suggested that the laziness, "dirt-eating" and general lack of industriousness attributed to the poor whites of the South was due to hookworm. The press, particularly the wire services, picked up the story, and the idea that the "germ of laziness" had been discovered made good newspaper copy.

Suddenly, hookworms were big news and everybody in the country seemed to be talking about these small roundworms. The news of the discovery of the germ of laziness also came to the attention of John D. Rockefeller. Rockefeller contributed a million dollars, a princely sum indeed in those days, to be used for the "eradication" of hookworm. With the money, the Rockefeller Sanitary Commission was established, and by 1910 the campaign was underway. The commission coordinated efforts among the hundreds of separate town, county, state, and federal agencies that had previously tried (or not tried) to do something about the hookworm problem, mostly with little success.

The campaign was a combination of mass medication and education. People lined up in country schoolhouses and similar places with mouths open to be medicated, and listened to lectures on how to prevent hookworm, often in the same schoolhouse where they were medicated. The basic message was "use the outhouse and wear shoes."

The Rockefeller anti-hookworm crusade was very successful. The incidence of hookworm in the southern United States was dramatically reduced. The campaign was also the model for other campaigns, such as the efforts to eradicate malaria conducted by the World Health Organization (WHO) of the United Nations. WHO is one of the few international organizations today in which there is true cooperation among nations, even among those who have been traditional enemies. Political differences among nations are frequently forgotten when the survival and good health of all, regardless of nationality, are at stake.

FACTS AND FIGURES

Hookworm is still around today, although it is not as serious a problem as it used to be. Estimates of worldwide infections vary, but about 700 million is a frequently offered estimate. The number sounds large, but many of these people have light infections that produce few, if any, symptoms. It still exists in the United States, but fewer than 50,000 Americans have the worms in sufficient quantities to cause problems.

Just how serious a problem hookworm is or will be in the

future is a matter of economics. Prosperous conditions mean good, high-protein food, and enough money left over to buy shoes. Prosperity also increases the possibilities of governments spending money on sewage works and education. Poverty and unsettled conditions brought on by war and other calamities are great for hookworms. In the United States, for example, hookworm increased during the years of the Great Depression. Many people could not afford shoes and the level of nutrition dropped.

There are a few places where hookworms seem to thrive in spite of prosperity. One of these is the tea-growing regions of Sri Lanka. The soil is of the right kind to support the hookworm larvae, and the tea plants provide protective shade. Many of the tea workers still prefer to work shoeless despite full knowledge of the existence of hookworm and how one gets it. This situation also exists among coffee pickers and other agricultural workers.

Treatment of hookworm infection is simple and effective; a wide variety of drugs is available. However, drug treatment is of little use if conditions that promote reinfection are allowed to prevail.

Trichina

In the discussion of pinworm, the statement was made that pinworm could well be the only parasite that most of us who live in industrialized, nontropical areas might ever get. There is, however, another strong candidate for that prize—*Trichinella spiralis*, commonly known as trichina. An important "if" clouds trichina's chances of being most likely to succeed in your interior. If you are an orthodox Jew, Muslim, or an observant follower of any other religion or group that forbids the eating of pork, your chances of ever getting trichina are virtually nonexistent. On the other hand, if pork is a regular inclusion in your diet, your chances of having been visited by at least a few trichina are excellent.

Trichina is the first parasite considered in this book that gains entry in food. It is also the first to have an intermediate host (of sorts) in its life cycle.

Trichina has been around a long time. Although the existence of trichina was not determined until around the 1830's, this worm was probably responsible for the following passage from Deuteronomy, the book of the Bible that contains much of the Mosaic law:

> . . . these ye shall not eat of them that only chew the cud, or of them that only have the hoof cloven: the camel, and the hare, and the rock-badger, because they chew the cud but part not the hoof, they are unclean unto you; and the swine because he parteth the hoof but cheweth not the cud, he is unclean unto you; of their flesh ye shall not eat; it is unclean unto you.
>
> Deuteronomy 14:7–8

While the proscription against pork kept the ancient Hebrews safe from trichina, the rules in Deuteronomy offered no protection against tapeworms obtained from eating fish or beef. Of course, the writers of the Bible did not have any experimental science at their command. What they knew was inferred from years of observations. They did quite well with the limited knowledge they had.

LIFE CYCLE

At this point it may be somewhat redundant to say that

trichina can be obtained from pork. Pork is indeed one of the many sources of this parasite. However, though there are many animals that harbor trichina, notably bears, the pig is the one of greatest direct importance to the possibility of human infection in the United States.

The life cycle of trichina is somewhat unique among human parasites, and it is difficult to find a good place to start explaining it. I'll start with the ingesting of encysted larvae by the host. Notice, I didn't say *eggs*, but live larvae enclosed in cysts in meat such as pork.

Digestive juices dissolve away the cysts and the worms emerge and move to the small intestine, where they grow into adult worms. Adult trichina are smaller than hookworms. The female is about 4 millimeters long and the male is less than 1½ millimeters in length. Males are thought to live less than a week, but females can live for two to three months.

The eggs of trichina hatch inside a uteruslike organ in the female. Fully hatched, active larvae emerge from the female. These very small (0.1 millimeter long) larvae do not leave the host, but bore into the wall of the intestine until they get into capillaries and are eventually carried into the blood circulation. The bloodstream carries them to voluntary muscles. These are muscles that we move at our will, such as muscles in the arms and legs. The worms, which by now are about 0.5 millimeter long, squirm around in the muscle fibers. The host, in a protective reaction, begins to lay down a covering composed mostly of calcium compounds on the worms. Eventually (about eight weeks usually), the larvae are completely covered over inside an oval-shaped cyst or capsule. The worms can remain alive, but inactive, in the cysts for one to two years.

A trichina larva in human muscle tissue. A cyst has not yet formed around the tightly coiled, tiny worm. It is, however, in the stage just prior to cyst formation. THE CENTER FOR DISEASE CONTROL

The only way for the life cycle to be continued is for the cysts to be eaten by another suitable host. Here the life cycle seems to reach a dead end. People are not, movies such as *Jaws* notwithstanding, the usual food of any other predator. Cannibalism is an extremely rare practice, and pigs just do not get much opportunity to eat people. It seems that trichina should have died out years ago, but trichina still thrives, although few parasitologists would say that it is a major health problem anywhere in the world. However, these same parasitologists will say that trichina *could* become a major problem at any time, as indeed it has from time to time, if we get careless and complacent.

If trichina comes to a "dead end" in human flesh, how does it keep going? Part of the answer is in human practices, the rest in nature. In many parts of the country pigs are fed garbage. While this practice has its advantages in terms of economics, it serves to keep trichina around. Garbage inevitably contains scraps of meat, and some of that meat is pork. Pigs eat the pork scraps and if the scraps contain trichina cysts, as indeed they might, the life cycle of trichina is continued. In some hog-growing areas, the garbage is cooked, much to the olfactory distress of those downwind from the cookers. While cooking does help to control the spread of trichina through garbage, it is impossible to kill all the trichina larvae. One ounce of pork can contain thousands of larvae.

Trichina is one of the most nondiscriminating parasitic worms known. It will establish in practically any mammal, and parasitologists have been able to get it going experimentally in a few birds. There are therefore vast reservoirs of trichina among wild animals. These animals pass trichina to each other through prey-predator relationships. One of the most important reservoirs is among rats. Rats are cannibalistic and readily spread trichina among themselves.

Insects that eat the carcasses of dead animals can act as temporary hosts. Any trichina cysts the insects might pick up from the flesh of the dead animals can stay alive for a few days inside the insect. If the insect is eaten by another mammal within that time, the mammal could get a few trichina worms along with its insect meal. Since pigs will eat almost anything, insects included, any insects that wander into a pigpen could serve as a minor mode of infection for pigs.

Undercooked or raw pork is the source of most of the world's human trichina infection. However, one of the highest known rates of trichina infection in the world occurs among people who rarely, if ever, eat pork. This high rate occurs among some Alaskan and Canadian Eskimos, about 25 percent of whom have trichina infections.

The Eskimos get trichina through a rather circuitous route. Whales and walruses are known to have trichina cysts in their flesh. When these animals die from natural causes, their bodies are eaten by scavengers, including sea gulls and crabs. The trichina cysts are able to survive in the birds and crabs. The birds serve to spread the cysts far and wide. The trichina cysts remain intact in the digestive tracts of the birds and come out of the gulls in their droppings. Crabs and some kinds of fish eat gull droppings and any trichina cysts that may be there are ingested, and they survive. Note that crabs have two ways of picking up trichina cysts in this complicated cycle: from dead whales or walruses and from bird droppings. If the fish and crabs are eaten in an undercooked or raw state, as they usually are by Eskimos, the larval worms will emerge from the cysts and begin anew in a human host. Trichina can also be picked up directly from eating whale and walrus meat. Polar bears and seal meat are other sources of infection. Practically all polar bears are infected.

Although the only way to get trichina is through the intermediate step of eating another animal, it is difficult to determine just which animal is the host and which is the intermediate host. In practically all parasitic life cycles that involve an intermediate host, the stage that lives in the intermediate host(s) is quite different in size and appearance from the form that exists in the primary host. The parasites also go through stages of development in the intermediate host, and these are different from those that take place in the primary host. This is not the case with trichina. The worms go through the same stages of development in a pig as they do in a human being; animals are the hosts or intermediate hosts for trichina only in relationship to each other. That is, as far as we are concerned, the pig is the intermediate host. However, as far as the pig is concerned, it is the primary host. The crabs, birds, and insects that pick up trichina cysts and pass them on are more like intermediate hosts. However, they are not true intermediate hosts either. The worms go through no changes in the insects, birds, and crabs. They are just carried. The relationship could be considered to be one of phoresis.

Scientific work on the life cycle of trichina began as the result of one of those fortuitous accidents that are legion in the history of science. In 1850, a German scientist named Herbst tossed a few pieces of dog and cat meat, left over from a dissection, to a caged badger. The badger died and, when Herbst examined the flesh of the dead animal, he found it to be loaded with trichina larvae. He then fed pieces of the badger meat to dogs and they too became heavily infected. It was not until ten years later that the adult *Trichinella* in the intestine were associated with the tiny larvae in the muscles. Prior to this time, a few "false starts" had led parasitologists to believe that the tiny worms in the muscles were the offspring of *Trichuris*. The relationship between adult trichina and the larvae was established by the German scientist, Rudolph Virchow, who fed meat loaded with trichina cysts to a dog. Adult trichina were then found in the intestine of the dead dog.

SYMPTOMS

Up to this point I have carefully avoided using the word "trichinosis." Among all the words used in discussing matters parasitological, trichinosis is one of the best known, and one of the most carelessly used. There definitely is a disease called trichinosis, and it is caused by all those trichina larvae boring around in the muscles. However, ingesting a few trichina larvae does not necessarily mean you will have the symptoms of this disease. Whether you have trichinosis is a matter of how many larvae are in the muscles at one time and how resistant the individual is to the worms.

Trichinosis is one of the most difficult diseases to diagnose. The symptoms are very much like many other diseases, including influenza. There are, undoubtedly, thousands of people who have had trichinosis and never knew it. The symptoms can begin with the establishment of the adult worms in the intestine and the release of larvae from the female. Symptoms of this first stage of trichinosis might include general internal distress or bellyache, along with nausea and diarrhea, or just a vague discomfort, or even no symptoms at all. Symptoms start about a week to two weeks after the cysts are ingested, and they are often dismissed as "some kind of virus" or, more colloquially, a "bug." Compared to what can happen next, they are minor symptoms indeed.

The second stage of the disease occurs when the larval worms are boring around in the muscles. This activity of the worms can cause extreme pain in the muscles and interfere with the ability to move the affected muscles. There is high fever and the victim can become noticeably swollen due to edema, the retention of fluids in the body tissues. This swelling is particularly noticeable around the eyes, so much so that it is regarded as almost a sure sign of trichinosis. There is also a great deal of sweating. This stage of the disease can last from three to five weeks. Sometimes the symptoms are so severe that the patient dies. If the patient survives the second stage (and most do), the third stage—the calcification and encapsulation of the cysts—begins. However, potentially fatal, second-stage symptoms can recur during the calcification process.

No one knows why the larvae in the muscles cause potentially deadly trichinosis in some people and a barely noticed discomfort in others. The number of worms is not the complete answer to that question, for some people have survived with worm loads twice as heavy as those that have killed other people. A rat can survive numbers of worms sufficient to kill hundreds of people. Rats have been known to survive loads of 22,500 larval worms per gram of muscle. Five larvae per gram or less (that's about 140 an ounce or almost 2,250 worms per pound of muscle) have been known to be fatal to human patients.

A person's ability to withstand trichina invasion is probably a matter of the individual's own body chemistry and the history of previous infections. As with many other diseases, a case of trichinosis will induce the body to form antibodies against the disease. Therefore, a few mild cases of trichinosis might be beneficial in the long run, in that they will protect the person against a massive invasion that might occur in the future.

TREATMENT

There is really no effective treatment for trichinosis, and much of the reason for this is the lack of dependable ways to diagnose the disease. Various drugs will dispatch the adult worms nicely, but usually by the time the presence of the adult worms is known, the larvae are well along boring their way into the muscles. Killing the adult worms then is almost, but not quite, the same as the proverbial closing of the barn door after the horse has been stolen. Expelling the worms from the intestine will prevent the release of more larvae to invade the muscles, but will do nothing about those that are already there.

The symptoms of the first stage of trichinosis are so much like other common human complaints that diagnosis is very difficult. There are ways to diagnose trichinosis in the second stage, which often resembles influenza, but these are not entirely dependable. A high level of a kind of white blood cell called the eosinophile (eosinophilia) is a good indication that trichinosis *might* be present. Eosinophilia is also a symptom of many other diseases, so it can only be suggestive, not conclusive, of the presence of trichina larvae in the muscles. Muscle biopsies are a more direct, if uncomfortable, way of diagnosis. A needle is inserted into the muscle and a little muscle tissue is drawn out. The tissue is examined with a microscope for the presence of larvae. A skin test, called the trichinellin test, is also used. A preparation made of powdered trichina worms is injected just under the skin. A positive reaction is seen in a bump or swelling at the injection site. The problem with this test is that it is likely to be positive if the patient has any inactive cysts from a previous infection.

About the only good that comes from diagnosing second-stage trichinosis is finding out that the affliction is not something else that can be effectively treated. There is not a great deal that can be done for trichinosis. It cannot be cured by any medical means. The patient's symptoms can be relieved to some degree with cortisone, to cut down the muscle inflammation, and by the administration of pain-killing drugs. The only thing that will bring about a cure is time and the patient's own body defenses.

FACTS AND FIGURES

There is much disagreement on just how serious a problem trichinosis is. It is undoubtedly one of the most widespread parasitic diseases in the industrialized, temperate zone nations. It is practically nonexistent in tropical areas.

One fact on which there is almost universal agreement is that trichinosis is not as serious a problem in the United States as it once was. In the late 1940's, estimates of the percentage of Americans infected with trichina to some degree ranged around 15 to 20 percent. Some 28 million cases of trichinosis were thought to exist in the world, and *75 percent* of these were thought to be in the United States! However, at the same time, the trichinosis problem in eastern Europe, particularly in Poland, was also quite severe.

Today, approximately four percent of Americans are thought to have trichinosis. However, when dealing with trichinosis, numbers can be deceptive. While the number of clinical cases of trichinosis—cases that require extensive treatment and/or hospitalization—is low (about 110 cases a year), some 150,000 to 300,000 trichina infections a year are believed to occur in the United States alone. No one knows, however, how many cases of trichinosis go undiagnosed. It is quite possible that thousands of mild and even severe cases of trichinosis have been treated as influenza, arthritis, measles, conjunctivitis, and lesser-known afflictions, such as periarteritis nodosa (an inflammation of the coatings of small and medium-sized arteries). Hundreds of thousands of low-level infections probably never come to the attention of physicians. These are instances in which there may be a sufficient number of worms to give the person a general "blah" feeling which he or she may dismiss with the usual "I must have some kind of bug."

The figures on how many cases there have been, the yearly incidence, and so on, are largely based on autopsies. While the cysts are too small to be seen without a microscope, the

calcified deposits over the cysts appear as fine, gritty particles. These particles were seen by anatomists hundreds of
years ago, long before anyone had any idea of what they
were.

Public health experts will continue to argue over how
much time and money should be spent on the trichinosis
problem. Some argue that no disease that results in only a
few hundred clinical cases, and no more than a handful of
deaths, a year can be considered to be a major health problem. There are far more serious problems on which public
health funds could be better spent, they say. On the other
hand, there are those who maintain that while the actual
number of serious cases a year is low, there is a constant
threat of outbreaks of trichinosis. Such outbreaks do indeed
occur from time to time, and not all of them involve pork.
Some of the most serious, such as one that occurred in
Greenland in 1947 and resulted in three hundred cases and
thirty-three deaths, originated in the whale-walrus-bird-crab
circuit discussed earlier. There are also occasional outbreaks
among hunters who eat undercooked meat from bears they
have killed. Eating undercooked bear meat is particularly
dangerous and can result in very heavy cases of infection.
Many of the fatal trichinosis cases on record involve bear
meat.

However spectacular the bear, whale, and walrus-related
outbreaks may be, all parasitologists agree that as far as
people are concerned the major source of trichina is pork.
From time to time, there are outbreaks stemming from
homemade sausage, particularly some ethnic varieties in
which the meat is smoked rather than cooked. The danger
is greatest if the pigs were also raised at home and fed on

uncooked garbage. Country preachers and schoolteachers were once frequent trichinosis victims. By virtue of their special position in the community, they were often given the honor of tasting and passing judgment on sausage in the making at country fairs and similar occasions. Another source of infection is hamburger that may have been stretched with pork scraps. One could get a surprise from consuming a rare hamburger prepared from such meat.

The chance of getting trichina from corn-fed pigs is low—about 0.1 percent of such hogs are infected. Laws requiring the cooking of garbage have reduced the incidence of trichina in that group of hogs to about 0.4 percent. Many public health experts say that until we know more about the trichina traffic among wild animals and how this traffic may overlap domesticated pigs, we cannot afford to be complacent.

In the United States pork is not inspected for trichina. For a while pork was examined with microscopes and an instrument called the trichinascope. These measures were found to be ineffective. Low levels of infection frequently got by the inspectors. Inspecting all pig carcasses for trichina would be very expensive and drive up the price of pork.

For all the arguments over what to do about the trichinosis problem, if indeed a problem exists, the prevention of trichinosis is really quite simple. All pork and pork products must be cooked until *all* the pink is gone and the meat is gray. Pork is one meat that should not be eaten rare under any circumstances. Freezing will kill trichina cysts; however, the meat must either be quick frozen at −37° C. or kept at temperatures of −15° C. or lower for twenty days. Even with this kind of chilly treatment there is still the possibility that some larvae deep within the meat may survive, so thorough cooking should still be the rule.

Strongyloides stercoralis

Strongyloides is a small roundworm. The female is about 2.2 millimeters in length and, as usual, the male is considerably smaller. It is found in much the same areas as hookworm, and that includes the southern United States. The World Health Organization estimate of its worldwide incidence is 35 million. The estimate for North America is 400,000.

LIFE CYCLE

The life cycle of *Strongyloides* is particularly interesting from a biological viewpoint. The worm exists in both free-living and parasitic forms, and continues its existence in the great outdoors and the small intestine through three distinct life cycles.

Life Cycle I: Direct

This life cycle is practically identical to that of hookworm. The worms live deeply embedded in the mucosa of the small intestine. They do not, as far as is known, suck blood. Eggs pass out in the feces and hatch into a stage that feeds in the soil. This stage lasts for two to three days and then the worm becomes an infective larva that can penetrate the skin as do hookworms. The journey of the worms through the lungs and so on is identical to that of hookworms.

Life Cycle II: Indirect or Free-Living

When the worms hatch from eggs passed in the feces of the

host, they develop into free-living adult worms rather than infective larvae. They reproduce more generations of worms that can, depending on the species, either become infective larvae or develop into another free-living generation. *S. stercoralis* usually goes through only one free-living generation that produces an infective generation.

Life Cycle III: Autoinfection

Worms hatch from eggs in the host's intestine and the larval worms bore into the small intestine, get into the bloodstream, go on a hookworm-type journey, and return to the small intestine to become adults. The larvae can also gain entry by penetrating the perianal skin as they emerge from the anus.

Strongyloides seems to have become adapted to deal with practically any situation. The free-living phase can continue for many generations in some species. The worm goes back to its parasitic ways when conditions in the outside world become difficult. Among the conditions that can trigger the change to the parasitic form are extreme cold and dryness.

SYMPTOMS

The disease, often referred to as "Cochin-China diarrhea," occurs in many degrees of severity. Light infections go unnoticed. Heavy infections, however, can result in a debilitating, potentially fatal condition. The ability of the worm to directly reinfect the host often results in extremely heavy infections.

Patients with moderately heavy infections might have abdominal pain, alternating diarrhea and constipation, and nausea. When the infection is really heavy, dysentery can become chronic, and the intestine might become so damaged that bits of the mucosa slough off and pass out of the body. The damage opens the way for bacterial infections, which can sometimes wreak more havoc than the worm does. Fatalities can occur in these extreme conditions.

TREATMENT

The same drugs used for hookworm, *Trichuris*, and many other intestinal nematodes are generally effective against *Strongyloides.* Side effects such as nausea and dizziness often occur. Other treatments may also be needed if the worm has done extensive damage to the intestinal mucosa and if secondary infections are present.

FACTS AND FIGURES

It seems that *Strongyloides* is going to be around for a long time. The autoinfection route reduces the effectiveness of sanitation measures, such as using toilets. There is also the possibility that dogs might serve as reservoirs.

The rate of infection among institutionalized individuals so severely retarded they cannot be taught to use a toilet approaches 100 percent. The fecal environment in which these people exist provides ideal conditions for the worms; the rate of reinfection is very high.

Filarial Worms

The rest of this chapter is devoted to filarial worms, which differ from the other nematodes discussed in this chapter in a number of ways. These worms do not live in the intestine of the host. They live in the blood and tissues and, as such, can be considered to be true endoparasites. The adult worms are very long and thin, whereas the immature worms are microscopic. Filarial worms have life cycles that involve stages in arthropod intermediate hosts, such as flies and mosquitoes. More than 250 million people have filarial worms, according to the World Health Organization. Numbers this large have a way of losing their meaning, but the personal tragedy of filarial diseases, some of which cause blindness and disfigurement, is very real to the victim.

Wuchereria bancrofti

Although *Wuchereria bancrofti* is hardly a household word, a word used to describe one of the more spectacular of the possible effects of harboring this worm is one that is fairly well known—elephantiasis. Most of us have seen pictures of people with grotesquely swollen limbs, breasts, genitals, and other body parts. Few people who shudder at the sight of this disfiguring disease know that it is caused by a thin, delicate, hairlike worm. Those of us who tend to be worried about these things will be relieved to know that *Wuchereria* is limited to tropical and subtropical regions. If you happen to be a resident of such a region you will be further relieved to know that elephantiasis is by no means an inevitable result of harboring *Wuchereria.*

LIFE CYCLE

The adult female is about 10 centimeters (4 inches) long and no more than 0.25 millimeter in diameter. The male is about half as big in length and girth. The worms live in the lymphatics, a complicated system of channels, vessels, and various structures that conduct fluid from between the cells back into the general blood circulation.

The females give birth to tiny young called *microfilariae.* The word "micro" means small, and the name fits. *Wuchereria* microfilariae are about 250 micrometers long and about eight micrometers in girth. They leave the lymphatics and make their way into the bloodstream.

Seeing microfilariae in the blood is one way to diagnose filariasis. However, you can't draw blood from a suspected host at any time. In most parts of the world where the parasite exists, the microfilariae exhibit a curious periodicity. They come out at night, usually between 10 P.M. and 3 A.M., give or take an hour. Sometimes you might be able to detect a few swimming around in the blood during the daylight hours, but the greatest concentration occurs during the night hours. The microfilariae retreat to the lungs when they are not moving around in the circulatory system. There is a variety that exists in the South Pacific that comes out into the blood between noon and 8 P.M.

The next stage in the life cycle occurs when some microfilariae are ingested by a mosquito while the insect is taking a blood meal from the host. Different species of mosquitoes are suitable intermediate hosts in different parts of the world. Once they get into the mosquito's stomach, the microfilariae immediately bore through the stomach wall and into the thoracic (chest) muscles of the mosquito.

The microfilariae go through a series of changes in the mosquito muscle. After about two weeks they become infective larvae, about one to two millimeters long. Microfilariae do the mosquito no good. A mosquito can handle only so many of them; more than one hundred will kill the insect. On the other hand, if there are fewer than fifteen or so microfilariae in the mosquito, the worms will not develop.

The infective larvae migrate from the thoracic muscles to

the proboscis, the mouth part of the mosquito that does the skin piercing and blood sucking. When the mosquito bites a host, the infective larvae come out of the proboscis onto the skin of the host. They get in through the tiny hole made by the mosquito's piercing proboscis.

Not too much is known about the details of the development of the parasite from the infective larva stage to the adult stage in the primary host. The development from larva to adults producing more microfilariae takes from six months to a year.

SYMPTOMS

The symptoms of filariasis vary from the inconsequential to the horrible. No proof exists that the microfilariae do any harm to the host, although it seems that large numbers of them congregated in the lungs would produce some pneumonialike symptoms. The only conditions in which microfilariae might be implicated are eosinophilia and granular growths in the spleen, neither of which are particularly serious situations. Ironically enough, the worms—both adults and microfilariae—can sometimes do more harm dead than alive. Microfilariae killed by drugs can bring on allergic reactions, and dead adults can cause some of the more horrifying results of filariasis. Parasitologists recognize three general categories of filariasis. There is, of course, no clear-cut division between them.

Asymptomatic filariasis

This disease is fairly common and occurs most often in people who are exposed as children. The hosts suffer no apparent ill effects other than a moderate swelling of lymph nodes here and there, and an eosinophilia. Microfilariae are found in fairly high numbers in the host's blood. Eventually, the adult worms die and, unless there is reinfection, the microfilariae disappear.

People who do not have allergic sensitivities to the dead and decaying worms suffer no ill effects from the worms' demise. Although these lucky people suffer no symptoms, many authorities feel they should still be treated since they serve as carriers of microfilariae. Often it is difficult to get people who are suffering no symptoms to agree to treatments that can make them nauseated, dizzy, and listless for a while.

Inflammatory filariasis

This is mainly an allergic reaction to the metabolic products of living worms and the breakdown products of dead ones. There is frequently extensive involvement of the male reproductive system, including inflammation of the spermatic ducts, testicles, and lymphatics in the genital area. Other organs, such as the spleen and kidneys, can be affected. Lymphatics of the arms and legs are also involved and there is often swelling, redness, and tenderness in the arms and legs. Headaches, chills and fever, vomiting, and just general poor health are frequent symptoms. Secondary infections from fungi and bacteria are fairly common.

Obstructive filariasis

When obstructive filariasis occurs, it is the final and often rather spectacular result of harboring *Wuchereria*. This is the disease commonly referred to as elephantiasis. Elephantiasis develops slowly after many repeated reinfections. Basically what happens is that as the worms die their decay products are absorbed. The worm is replaced by what is essentially scar tissue. There is massive enlargement of lymph nodes and inflammation of lymphatic vessels. The flow of lymph is blocked, and the accumulating fluid causes swelling. The elephantlike appearance of limbs is caused by the growth of excessive connective tissue stimulated by the high protein content of the lymph. Much of connective tissue is collagen, a protein. The tissue in the swollen body parts hardens over a period of years into the characteristic elephant-skin appearance. Commonly affected areas are the legs, arms, breasts, scrotum, and external female genitalia.

Before we go on, it should be emphasized that the tree-trunk-sized legs and bushel-basket-sized scrotums often seen in the nightmarish pictures of elephantiasis victims are extreme cases and are very much the exception. The incidence of obstructive filariasis varies from less than one percent in

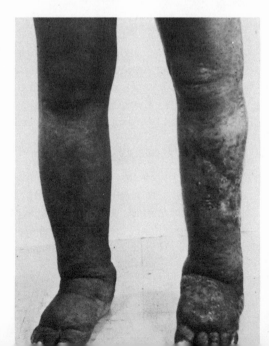

A case of elephantiasis with involvement of both legs
THE CENTER FOR DISEASE CONTROL

some areas to more than 70 percent in others. However, even in those areas where the incidence is high, extreme examples so often seen in pictures are quite rare.

TREATMENT

Diagnosis is made by microscopically examining blood that has been drawn at night for the presence of microfilariae. Diagnosis of all microfilarial infections is made in the same way. Identification of the particular species involved is made by staining the slides and examining the microfilariae for identifying characteristics. If obstructive filariasis occurs, there are usually no microfilariae in the blood. However, the presence of elephantiasis is obvious and no sophisticated diagnostic techniques are required.

A drug called diethylcarbamazine (sold under the name Hetrazan) does a very effective job of killing microfilariae. The drug also either kills adult females or renders them sterile. Some bothersome side effects, including nausea, dizziness, headache, and fever, often occur. However, these side effects may be due as much to the death and disintegration of worms as to any effect of the drug.

The effects of inflammatory and obstructive filariasis are also treatable. Pressure bandaging helps to control swelling. Steroid drugs help to reduce inflammation and swelling to some degree. Surgical repair of organs affected by elephantiasis is often helpful. The results of surgery depend on how badly the body part is affected and the general condition of the patient.

FACTS AND FIGURES

Wuchereria is a widespread parasite in tropical areas of the world. It is found throughout most of Southeast Asia, including southern India, Burma, Thailand, Sri Lanka, Vietnam, Cambodia, Laos, Malaysia, Indonesia, the Philippines, and New Guinea. It occurs also in coastal China, Taiwan, northern Australia and in southernmost Japan and Korea, and in most of the Pacific islands comprising Micronesia and Melanesia. It also afflicts people in central Africa and in the Western Hemisphere in the Caribbean islands, coastal South America (Caribbean and Atlantic), and Central America. *Wuchereria* may have been brought to the Western Hemisphere by the slave trade. It is known that the disease existed in the United States, in the Charleston, South Carolina, area, until about sixty years ago. That infection was definitely imported with slaves.

Although seen occasionally as far north as Spain, *Wuchereria* is primarily a tropical parasite. Europeans and Americans, however, have been exposed to it through hundreds of years of colonial and military ventures in tropical areas. Many of the soldiers, sailors, and marines who participated in military actions in the tropics, from Captain Cook's eighteenth-century explorations right on down to the Vietnam War, were more afraid of elephantiasis than they were of enemy spears or bullets.

The fear of elephantiasis did produce a few benefits. Commanders of troops in tropical areas were rightfully more concerned with malaria than with elephantiasis. Tens of thousands of American military personnel who served in the tropics in World War II and Vietnam contracted malaria. One reason for the high incidence was that the men tended to get careless about taking the necessary anti-mosquito measures. Exhibiting elephantiasis victims, particularly those with scrotum involvement, to troops, and informing them that the same mosquitoes that carried malaria also carried the "germ" that causes elephantiasis, made the men very conscientious about mosquito control.

Wuchereria can be controlled by killing mosquitoes and by mass treatment of people who harbor microfilariae. As pointed out earlier, some very effective anti-microfilarial drugs are available. In those areas where governments have spent money on control programs, the incidence of filariasis has gone down dramatically. The disease is almost absent today in Tahiti and the Virgin Islands, for example.

Environmental problems with insecticides used to kill mosquitoes have complicated control programs in many areas. However, mass use of drugs to kill microfilariae is still effective in these areas and is generally cheaper than netting, mosquito repellents, screens, and other mosquito barriers.

Onchocerca volvulus

Although slave traders could hardly be called kind and benevolent, they were concerned, to some degree, with the health of their human merchandise. Dead slaves could not be sold and many slaves, even the healthiest, died during the voyage from Africa to America. Though traders expected to lose a certain percentage in transit, they wanted to minimize their losses as much as possible, so they learned to recognize certain signs of illness in the human bodies they bought and sold. Among these signs were roundish swellings on various parts of the body. A slaver was likely to reject an African with these swellings, or at least use them as levers to beat down the price. The slavers certainly did not know that these bumps or nodules were caused by a filarial worm called *Onchocerca volvulus,* but they did know that captives with these swellings were likely to get very sick, if they weren't already so, and that they might also go blind. Many slaves harboring *Onchocerca* were brought to the Western Hemisphere, and today this parasite is still a problem on the American continents in some areas of Mexico and Central and South America.

LIFE CYCLE

The adult worms live in the swellings or nodules the slavers looked for hundreds of years ago. Found mostly on the surface, the nodules also occur in deeper parts of the body. Females can get to be about 50 centimeters (about 20 inches) long and 0.5 millimeter in diameter. The males are much smaller, seldom attaining a length of more than 5 centimeters (about 2 inches).

The microfilariae migrate out of the nodules into the surrounding tissues just beneath the skin. Sometimes, but not very often, *Onchocerca* microfilariae can be found in the bloodstream and in various organs, particularly the eyes.

The intermediate hosts are species of black flies. These are bloodsucking flies capable of giving viciously painful bites. One of the principal intermediate hosts is a black fly with the rather appropriate name of *Simulium damnosum*. When the fly bites an infected host, microfilariae are taken up into the fly's stomach. The microfilariae break into the fly's thoracic muscles and develop into infective larvae that move into the fly's proboscis. The next time the fly bites, the larvae move onto the skin and get into the host through the bite hole left by the fly.

As the larvae mature into adults, they wander around in the tissues just under the skin. They gather in groups and are encapsulated in the nodules referred to earlier. The nodules are formed by the host in an allergic reaction to the worms. The worms exist in the nodules; if the nodules are surgically opened the worms are revealed, but it is difficult to tell where one worm ends and another begins. The nodules are from five to twenty-five millimeters in diameter.

The nodules have different patterns of location in different parts of the world. In Africa they tend to be on the trunk, arms, and legs. In Mexico and Central America they are more often on the head and are less likely to be on the trunk and limbs. In Venezuela, however, the nodules seem to follow the African pattern.

Just why these patterns exist has long been a favorite topic of discussion among parasitologists. Some have proposed that the African and Western black flies are of different strains and have different behavior patterns. Others have offered a more simple explanation. People in Mexico and Central America generally wear clothing that covers most of the body, leaving the head as the body part most accessible to the flies. The black flies live mostly in the cooler mountain regions—coffee-growing country. This area is much cooler than the lowlands. Most of the women who work in the coffee fields wear ankle-length dresses, leaving only the head, neck, and arms exposed. Africans, particularly in the hot, humid equatorial inland regions, tend to wear less clothing, thereby giving the flies a greater range of target. As industrialization spreads through Africa, the old cultural patterns, including those involving clothing, are changing. It will be interesting to see what effect, if any, these changes will have on the distribution and frequency of onchocercal nodules.

SYMPTOMS

The general health of the skin can be affected by *Onchocerca.* Most of the damage is caused by the microfilariae, rather than by the adults. The skin loses elasticity, thickens, and may lose its pigment. A frequent effect of loss of elasticity is so-called "hanging groin." In this condition, the skin on the upper thigh loses so much elasticity that it hangs in a kind of pouch near the genital area. The skin can become hot, swollen, painful, and itchy. These conditions come and go and, over a period of time, the skin thickens. Excessive wrinkling of the skin might also occur. In Africa there is sometimes a patchy loss of pigment in the skin, resulting in a condition called "leopard skin."

The most serious of the possible results of onchocercal disease is blindness. Microfilariae wander through the body and many become trapped in the eyes. They congregate in the front part of the eye in the cornea, choroid, and iris, the part of the eye that gives it its color. They also congregate in the vitreous body, the jellylike material in the interior of the eyeball. The effect of the microfilariae is slow but steady. White, opaque patches begin to form in the cornea, the transparent part of the eye that allows light to enter. The appearance of these opaque patches starts generally around seven years after exposure to the parasite. The next involvement can be an irritation of the iris. The iris thickens and becomes irregularly shaped. The lens may become fused to the iris. There may also be damage to the optic nerves. Just how much vision is lost is a matter of how far the opaqueing of the cornea and the degeneration of the iris will proceed.

These changes that take place in the eye stem from toxic products given off by dead and living microfilariae, toxins from adult worms, and allergic reactions on the part of the host.

TREATMENT

Surgical removal of the nodules, with the worms therein, is a direct and usually effective procedure. The fewer adult worms there are, the fewer microfilariae there are likely to be wandering around. However, it is a useless procedure if large numbers of microfilariae are already in the eyes. It is also difficult to find all the nodules, particularly in Africa, where some are so small they are hardly noticeable.

Drug treatments that kill the microfilariae are also available, but they have to be used carefully. Allergic reactions to all those dead microfilariae can be quite severe in some people. Drug treatment, however, is often necessary to stop the effects of the disease. If treatment is started early enough, the degeneration of the eye can be stopped and often reversed.

FACTS AND FIGURES

The World Health Organization estimate for the incidence of onchocerciasis is over 20 million. Fortunately, only a small percentage of the people with onchocerciasis suffer from blindness. The rate of blindness varies in different areas. For example, in rain forests of West Africa, the rate is about 2 percent. In savanna areas, fewer people have onchocerciasis, but a greater percentage of them, about 5 percent, become blind from the disease. The blindness rate is about 8 percent in Central America.

Onchocerciasis is found mostly in the vicinity of fast-moving, small streams. For that reason it is called "river blindness" in some places. The black fly intermediate hosts lay their eggs only in these small streams. The eggs are deposited on stones, wet leaves, and other debris in the stream. Larvae hatch from the eggs and spin open-ended cocoons that are attached to the rocks with fine silk threads. The current brings food to the larvae. Although they are capable of flying for great distances, black flies seldom fly more than ten miles or so from where they hatch. This concentration of black flies around rivers and streams also means a concentration of onchocerciasis around these bodies of water.

The disease has been brought under control to some degree in Central America. Mass treatment, involving removal of worms and use of drugs to kill the microfilariae, helps to break the circuit in the life cycle. Stopping the life cycle at the human point is much easier than trying to kill large numbers of black flies. These control measures have not been as effective in Africa. Africans tend to have more nodules because they wear less clothing than people living in the cooler highland areas of Central America. Control of black flies through insecticides and altering streams to reduce the speed of the current has proved to be effective in some parts of Africa. It is unlikely, however, that black flies will ever be eradicated, and this means that onchocerciasis will probably continue to afflict many people, although the overall incidence of the disease should continue to decrease.

Loa loa

This worm is much smaller than the other filarial worms discussed here, and it usually does far less damage to the host than the other filarial worms that commonly infect people. Although mainly a human parasite, it has also been found in monkeys. The females are about 5 to 7 centimeters (2 to 3 inches) long, and the males 2 to 3.5 centimeters (0.8 to 1.4 inches) in length. The most notable thing about this worm is the startling way in which it can make its presence known. The adult worm is a wanderer, and sometimes can be seen whipping through the eye in the thin membrane called the conjunctiva.

LIFE CYCLE

The life cycle is similar to that of other filarial worms. The adult worms live in the subcutaneous tissues, producing microfilariae as they go. The young ones get into the bloodstream, where they are taken up into the intermediate host, the mango fly (*Chrysops*), when it bites. The large mouth parts of the mango fly can inflict a very painful bite. The development of the microfilariae into infective larvae, and the way the microfilariae get into the host from the fly's bite, follow the same pattern as that of *Wuchereria* and *Onchocerca*.

SYMPTOMS

The worms develop into adults and wander through the body in the subcutaneous tissues. From time to time, swellings (called Calabar swellings) may appear in various parts of the body. Generally several inches in diameter, these swellings can be itchy and painful. They are probably allergic reactions to dead worms and the metabolic wastes of living ones. Sometimes worms are found in the swellings and sometimes not. *Loa loa* is a particularly long-lived worm—some live as long as seventeen years—and the Calabar swellings may recur as long as there are any worms, dead or alive, in the host.

The worms generally do not cause any discomfort in their wanderings, except when they wander across the eye and the bridge of the nose, which really hurts. However, the startling and painful appearance of the worm in the eye offers an opportunity to extract the worm (by a qualified physician). The extraction must be done quickly, since the worm is a fast mover and dashes off to the safety of the host's interior if it is disturbed during its ocular sojourn.

TREATMENT

Loa loa infection can be successfully treated with drugs and by surgical removal of the worms.

FACTS AND FIGURES

Loa loa is found only in Africa, particularly in west Africa, the Congo River basin, and the southern parts of the Sudan Republic. However, doctors in the United States and in other countries to which *Loa loa* is not native sometimes get a chance to see it in the eyes of students from African nations who are studying at American universities.

Of some concern to health officials is the knowledge that *Loa loa* microfilariae can develop into infective larvae in an American insect, the American deerfly. There are, however, no indications that this parasite will ever become established in the United States.

Dracunculus medinensis

> And the Lord sent fiery serpents among the people, and they bit the people; and much people of Israel died.
> And the people came to Moses and said: 'We have sinned because we have spoken against the Lord, and against thee; pray unto the Lord that He take away the serpents from us.'
> And Moses prayed for his people.

And the Lord said unto Moses: 'Make thee a fiery serpent and set it upon a pole; and it came to pass that if a serpent had bitten any man, when he looked upon the serpent of brass he lived.

Numbers 21:6–8

Dracunculus, known the world over as the guinea worm, long provided Europeans and other Westerners with one of the more exotic sights of their trips to the Near East and India. Visitors were astonished to see many of the local inhabitants walking around with sticks or twigs attached to various parts of their legs with what appeared to be a white string that issued from *inside the leg.* Natives were also seen slowly and laboriously wrapping more of this "string" from inside their legs around the twig. All of this may have been exotic fare to the Europeans, but it was part of the day-to-day routine for the local people, who were employing a time-honored method of ridding themselves of guinea worms.

LIFE CYCLE

Adult guinea worms live in the cutaneous and subcutaneous tissues—areas just below the skin. They are found in the subcutaneous areas of the arms, legs, shoulders, and trunk. Generally, only female worms are found. No one is quite sure what happens to the males. Sometimes the presence of wandering worms can be detected as a moving ridge, just under the skin, somewhat like a rather mobile varicose vein. The female is a slender lady, no more than 1.7 millimeters in diameter. What she lacks in girth, she more than makes up for in length. They can be from 50 to 120 centimeters long (from 20 to 48 inches). Although guinea worms have much in common with filarial worms, they are not considered to be true filarias.

The female guinea worm sends her young out into the world in a rather direct fashion. By the time the worm has reached sexual maturity, she is literally filled with tiny larvae. At this time she moves toward the surface of the skin, frequently on a wrist, ankle, or foot. She secretes enzymes that break down some of the tissues under the skin, and pushes through the deep layers of skin. This action causes a blister to form. When the blister breaks, an ulcer is formed and a little hole in the skin is exposed. The worm pokes her head end out through this hole.

When the protruding end of the worm comes into contact with water, a uteruslike organ of the worm bursts and thousands of larval worms in a milky fluid are released into the water. After each ejection, the open end of the uterus dries up and seals shut. Another contact with water will cause more larvae to be ejected. When all the larvae are ejected, the female worm dies and is absorbed by the tissues.

The tiny larval worms swim about in the water. If they are not eaten by a cyclops in a few days they will perish; if they are, the life cycle continues. A cyclops is a tiny shrimplike organism commonly found in fresh water. It is called a cyclops because of the single eye it bears at its anterior end.

When the larval worms are eaten they are not digested; instead, they burrow into the body cavity of the cyclops where, in about three weeks, they change into the infective form. The worms get into a new host when and if the cyclops are ingested, usually in drinking water. The larvae break out of the cyclops as the tiny shrimp are digested by the host. They burrow into the intestinal wall, get into the bloodstream, and are carried to connective tissue where they mature. Mating is believed to take place during the migration. Males are seldom seen in the subcutaneous regions. Scientific proof of the existence of male guinea worms was not obtained until the 1930's.

The life cycle of *Dracunculus* is evolution at its blind best. The way in which this worm has become adapted to the habits of its hosts and intermediate hosts is amazing to behold. Of course the victim might not think it is so magnificent, and the worm doesn't think about it at all; it just does what its adaptations direct it to do.

SYMPTOMS

The worms produce no noticeable symptoms when they are migrating through the body toward the subcutaneous regions. The blisters and ulcers that form when the worms carry out their motherly duties can be bothersome. In many hosts there are a variety of symptoms just before the worm breaks through at the blister site. These symptoms include itching, burning sensations, difficulty in breathing, and vomiting. Most of these symptoms are believed to be allergic reactions.

Sometimes the worms die before they reach the subcutaneous regions. When this happens they may disintegrate, become absorbed, or become calcified. Any of these possibilities can bring on allergic reactions and the calcification can mimic symptoms of many other diseases.

TREATMENT

The best treatment is still one of the oldest, and that is removal of the worms. Of course, the techniques of removal employed today are a bit more refined than the traditional method, but the expected result is the same. This method, which so astonished Europeans when they first witnessed it years ago, involves irrigating or douching the ulcer with water. This induces the worm to protrude a bit more from the hole she has made in the skin. The protruding end of the worm is then carefully wrapped around a stick or twig. The process is repeated and each time a little more of the worm is wrapped around the stick. A successful extraction takes about two weeks, a not surprising length of time when one realizes that the other end of the worm might be as high up as the hip. Extreme care and patience are required for this delicate task. If the worm is pulled too hard it will break, and that will bring on all kinds of trouble. The worm will retract into the tissues and disintegrate. The products of the worm's decay are very injurious to the surrounding tissue. Bacterial infections inevitably follow, and gangrene is also a frequent result. Loss of the leg following such an accident is not uncommon.

The biblical instruction to "Make thee a fiery serpent and set it upon a pole" may be an allegorical allusion to the twist-on-a-stick removal method. The staff of Aesculapius, the representation of a serpent around a stick that is often used as a symbol of medicine, could have its origins in this ancient treatment. The description of "fiery" probably came from the burning and itching pain caused by the worm's activities. The scientific name of the worm, by the way, means dragon of Medina.

Surgical removal of the worms, with the help of antibiotics and sterile techniques, is a widely used treatment today. Various drugs, such as Thiabendazole, have been known to cause the worms to be expelled spontaneously or, if not that,

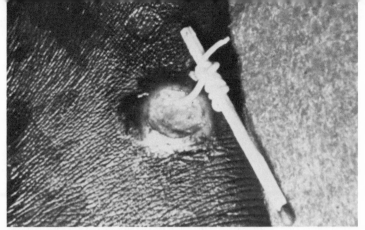

The stick that looks like a wooden match is called an orange stick. Its usual use is in the care of fingernails, but in this picture, the stick is being used for a medical, rather than cosmetic, reason. A female guinea worm can be seen extending from the ulcer in the host's foot (the heel is to the left). About a foot of worm has been wrapped around the stick, but there is still quite a bit of worm to go before all of it is removed. **THE CENTER FOR DISEASE CONTROL**

to weaken or kill the worms so that removal by a refinement of the stick method or by surgery is easier.

FACTS AND FIGURES

Dracunculus is fairly widespread in the world's tropical areas. It is found in much of Asia, from the Arabian peninsula to India. In Africa it is prevalent in the Nile Valley on up to the Sudan. There are scattered areas of incidence through equatorial Africa and in West Africa. Indonesia is another area where it occurs. It is sometimes found as far north as the southeastern Soviet Union. *Dracunculus* is another parasite that was brought to the Western Hemisphere by the slave trade. Although the disease died out in the Western Hemisphere years ago, reservoirs of *Dracunculus* exist in dogs and other carnivores. Raccoons, as a consequence of dipping their food in cyclops-laden water, harbor a similar species. There are occasional human cases in North and South America that stem from contact with these reservoir hosts.

One of the great paradoxes of parasitology is that the guinea worm, an organism that must have water to continue its species, is found in some of the driest parts of the world. However, people who live in desert areas tend to congregate where there is water, whether around an oasis pond, a stream, or a river. The guinea worm does quite well where there are concentrations of people in the same areas where the density of the intermediate host is high.

The World Health Organization estimates that as many as 80 million people may have *Dracunculus.* Education of people who are in danger of infection is one way to cut down on the incidence of guinea worm infections. However, some customs and religious practices, such as frequent ablution (ritual cleansing in rivers and other bodies of water), help to keep the waters richly supplied with larvae. In many parts of the world, people with active *Dracunculus* ulcers bathe or wash clothes in bodies of water that serve as public water supplies. Although water from a tap seldom contains cyclops, water taken directly from rivers is usually filled with them. Wells may also be contaminated with cyclops. Filtration and/or boiling of the water would eliminate the cyclops, but this is seldom done in many of the areas where *Dracunculus* is a problem. Fuel to boil the water might be too expensive, and the materials needed for filtering might not be available in many areas. *Dracunculus*, as is true of almost all other roundworms, are not about to leave the human scene for a long, long time—if ever.

5.

Tapeworms

"You put a saucer of milk on the table, open your mouth, and the tapeworm will come out to get at the milk."

"That guy is so skinny he must have a tapeworm."

"I've never seen anyone eat so much. She must have a tapeworm."

"Tapeworm" is what comes to mind when most people hear the word "parasite." As is the case with most things about which there is general familiarity, but not too much specific knowledge, there is quite a bit of folklore on tapeworms. The quotes that start this chapter are part of that folklore.

All the statements represent misinformation. A tapeworm is incapable of crawling out of anyone's mouth. They are strictly other-end emergers. The "crawl out of your mouth" myth probably grew out of the very real, occasional incidents of ascarids coming out of mouths and noses. Tapeworms do not necessarily make the host skinny and ravenous. Most people who harbor tapeworms retain whatever grade of appetite they had before taking on the tapeworm. Others are affected in such a way that they lose appetite, while others might feel some increased hunger due to the presence of the tapeworm. Tapeworm hosts show the same distribution of corpulence and leanness as the general population. One tapeworm host I knew was a healthy-looking two-hundred-pounder.

Tapeworms belong to the phylum Platyhelminthes, the flatworms. The class that includes the tapeworms is called Cestoidea. There are two subclasses: Cestodaria and Cestoda. The former, mostly parasites of fish, are not considered to be true tapeworms. The cestodes are the true tapeworms. All cestodes are parasites. In the adult stage they are, with few exceptions, internal parasites. Many of the larval stages exist as tissue parasites.

The tapeworm has been described by many observers as the ultimate parasite. The cestodes have probably existed for quite some time, and in that time have evolved advantageous life cycles. The evolution of their physical structure can be thought of as degenerate.

Tapeworms have few of the organs and systems of organs that other animals have. They don't need them. The host does it all for them. For example, tapeworms do not have a digestive system, nor do they have respiratory or circulatory systems. They do have ways of getting rid of metabolic wastes, and there is a primitive nervous system. They are capable of very limited movement.

Most tapeworms cannot be thought of as one individual. The very nature of most of these parasites is foreign to the idea most of us have of what an individual living thing is.

If you saw a tapeworm you would see what looks like a flat, white ribbon. You would not, however, see a smooth, gift-wrap-type ribbon. You would see a ribbon made of many sections. It would appear to be "choppy" or serrated at the sides. The sections are not all of the same size. One end of the worm would be tiny, almost invisible to the unaided eye, and you would see that the farther away from the tiny end the sections are, the larger they are.

The basic structure of a tapeworm is a head followed by a neck region, followed in turn by series of sections. The words "head" and "neck" are used in a rather loose sense. The head is so called in the sense that it is at the end of the animal designated as the anterior or front end. There are no mouth, eyes, or other structures commonly thought of as belonging in a head. There is, in some species, a concentration of nerve tissue that could be thought of in some very remote way as the animal's brain. The sole function of the head is to latch on and hold on to the host's intestine. The heads of different species are equipped with a variety of hooks, sucker disks, and spines to aid in this function. The word *scolex* is a far better one to use when referring to a tapeworm's head.

The sections grow from the neck. As a new section forms, it pushes the one ahead of it down the line. The sections get gradually larger as they progress from the scolex. The immature sections are the ones nearest the neck. The degree of maturity increases as one goes down the chain of sections. The sections at the very end of the worm are filled with eggs, most of which are fertilized if all has gone well in the unplanned life of the tapeworm. Between the head and the last section, the sections are in every stage of development, ranging from infancy through childhood, adolescence, and maturity to the final stage, which is better referred to as "gravid" rather than old age. A gravid section is filled with eggs and ready to go out into the world and keep the species going. In most species, well over 90 percent of the area in a mature tapeworm is devoted to reproductive organs. To describe a tapeworm as a head followed by a string of reproductive organs is not an inaccurate description.

Tapeworms range in length from three-section worms, an inch or two long, to "monsters" of more than three thousand sections, forty feet or more long. Tapeworm sections are not the same thing as segments of an earthworm, and are properly called *proglottids.*

In most species each proglottid contains a full set of male and female reproductive organs. A tapeworm, or rather each proglottid, is a hermaphrodite. Self-fertilization within a section can occur, as well as cross-fertilization between different sections of the same worm or between sections of different worms in the same host.

Each proglottid is essentially an individual organism. There is, however, a cuticle that covers the entire worm. The sections are connected by a double nerve cord, bands of muscle, and excretory tubes. Since tapeworms are parasites of the digestive system, the proglottids and eggs pass out of the host in the intestinal wastes. Life cycles include those that involve no intermediate host and those that involve one or more intermediate hosts.

Almost all cestodes spend their lives in the host's intestine. A few live in the bile ducts and other parts of the digestive system. A tapeworm has no mouth or digestive system. It literally soaks up the digested food in the host's intestine. There is evidence that some tapeworms may go beyond just soaking up semidigested food: Some may act like living sandpaper and abrade the inner surface of the host's intestine.

The abrading probably releases fluids, high in protein, for the tapeworms to soak up.

There is very little oxygen inside an intestine and the ability to live in an environment so lacking in oxygen is an important adaptation of tapeworms and other parasites.

Few vertebrate animals are not subject to invasion by at least one species of tapeworm. There are tapeworms that commonly infect all of our domestic and pet animals. Some half dozen can become established in human hosts. All the usual tapeworms of humans, except one, have life cycles involving one or more intermediate hosts. Humans can commonly serve as the "intermediate host" of a few tapeworms. These immature forms of the tapeworms, living as tissue parasites in man, can be very dangerous.

Three common human tapeworms slip into our interiors in food. The first of those we will consider is the beef tapeworm.

Taenia saginata (or *Taeniarhynchus saginatus*)

One morning while I was teaching a high-school biology class, I received an unexpected visitor—the school nurse. Without so much as a knock on the door, she walked into the classroom and whispered a question in my ear.

"What's white, wiggles, and comes out in your stool?" she asked in an agitated whisper.

"Sounds like a worm," I whispered back.

"What kind of worm?" she persisted.

"There are all kinds of worms," I retorted, getting a bit annoyed as the class started to get restless. "I'd have to see it."

The nurse left and I went back to teaching, wondering why

anyone would ask me a question like that, and why she couldn't have waited until the class was over before disturbing it with weird questions. About five minutes later she was back. Without saying a word she handed me a piece of toilet tissue. On the tissue there was a flat, white object about three-quarters of an inch long and a quarter of an inch wide. The object was moving actively, contracting and expanding so vigorously that it did a couple of flip-flops. I soon forgot my annoyance as I realized that for the first time in my life I was looking at a living human tapeworm proglottid.

I forgot there were twenty-five wide-eyed, curious ninth graders in the room and blurted out, "Where did you get this? This is fantastic. Do you know what this is? It's a—."

She stopped me with a raised hand and whispered, "I'll tell you later."

Of course the class wanted to know what was going on. I stopped their clamorous questioning by telling them it was school business and none of their concern, but although that quieted them down, it did not satisfy them. They were still curious. It isn't every day you see the school nurse hand your teacher a piece of toilet paper.

I had worked in a veterinary hospital when I was in college and I had seen plenty of tapeworms and roundworms that had been expelled from cats and dogs in worming procedures or removed in autopsies. Actively moving sections of a dog tapeworm were frequent sights in the veterinary hospital. Beef tapeworm sections are active movers. I knew, however, that this proglottid had not come from a dog. The nurse had said "your stool" when she asked the question, not dog or cat stool.

Later I found out how the nurse had obtained that interesting specimen. Earlier that day, some of the teachers taking a break in the teachers' lounge heard screams coming from the adjoining ladies' room. All the women in the lounge rushed in to see what was the matter. The men teachers who were waiting anxiously outside the ladies' room door could make out something that sounded like, "Help! There's wiggling white spaghetti coming out of me! It's alive!"

The agitated lady was one of the school secretaries. By the time I spoke to her later in the day she had calmed down considerably. I asked her if she had ever eaten any beef obtained from a source other than the usual supermarket or butcher shop, and if she liked her beef rare. She had to think for a while, but then remembered that some two years earlier they had been given a gift of a worn-out dairy cow. They had slaughtered the animal and were thus provided with some free, but rather tough and stringy, beef. She went on to say that she was indeed addicted to rare beef. "I like it when it's practically still saying 'moo,' " she confessed.

I offered the opinion that the wiggling white spaghetti that had so disturbed the lady was a beef tapeworm. This diagnosis was later confirmed by laboratory tests. The women and other members of her family who liked their beef rare were treated and forever parted from their ribbony guests.

LIFE CYCLE

The story of the school secretary almost sums up the beef tapeworm life cycle. Adult *Taenias*, exclusively human parasites, are from four to six meters (about thirteen to twenty feet) long and contain from 1,000 to 2,000 proglottids. The head has four suction-cup-type disks but, unlike some other human tapeworms, it has no hooks. They live in the part of the intestine called the *jejunum*. The jejunum is the upper part of the small intestine. Gravid proglottids pass out in the stool, usually one at a time.

If the proglottids reach the ground, rather than a toilet, the life cycle has a chance of continuing. The active movement of the proglottid serves to expel the eggs. A single section can contain some 100,000 eggs. The next stage occurs when a cow, in the course of eating grass, swallows some of the eggs. The young tapeworm hatches out in the cow's digestive tract, bores into the intestine, and is carried in the blood to the cow's muscles, the part we call meat. The larva develops into a form called a *bladder worm* or *cysticercus* or, more simply, a *cyst*. Cysticerci are pinkish, oval-shaped structures about 9 millimeters (not quite three eighths of an inch) long. The cyst contains an invaginated sort of "upside-down" neck and scolex.

The life cycle is continued if beef containing the cysts is eaten by the primary host, a human being. Of course, the larval worm in the cyst must be alive if it is to grow into an adult tapeworm. In order to be alive the beef must be either raw or rare. Cooking at 57° C. (135° F.) will kill the cysts.

When swallowed by a human, the scolex and neck in the cyst turn right side out and proceed to grow into a tapeworm. Sections will start to appear in the stool eight to ten weeks after the bladder worms are ingested.

SYMPTOMS

Many, if not most, people who have beef tapeworms never know it. Few hosts suffer any adverse effects. Some people complain of gastric distress, nervousness, diarrhea, nausea, dizziness, or change in appetite. Interestingly enough, most of the complaints come from people who know they have the

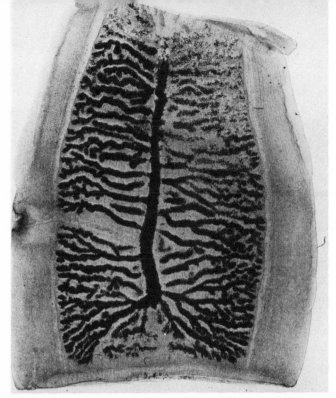

A photomicrograph of a gravid proglottid of Taenia saginata, *the beef tapeworm. This proglottid has been cleared, that is, treated with chemicals to make the internal structure more visible. The various branched structures comprise the uterus, the organ that contains the eggs.* **THE CENTER FOR DISEASE CONTROL**

tapeworm. Those who don't know seem less bothered by such symptoms, if they have them at all.

Most people find out about their guests the way the school secretary found out. They see the sections on the stool or in underwear. As the story of the secretary suggests, knowledge of the presence of a tapeworm so gained can be disturbing. Knowing about a beef tapeworm frequently is more disruptive of the host's health and tranquillity than anything the tapeworm ever does.

The beef tapeworm, however, is capable of doing a few things that are, if nothing else, surprising and disconcerting. The active, vigorous *T. saginata* proglottids often do not wait for the host to go to the bathroom to make their debut in the outside world. Frequently, they break off and emerge from the anus on their own power. The host can often feel that something is going on there and, if there is no knowledge of the tapeworm's presence, the host might think an involuntary bowel movement has occurred. Fortunately, *T. saginata* emerge one at a time. However, on those few occasions when several of them come out at one time, either individually or in a string, the host could be thrown into a panic. Such was the case with the school secretary. Some hosts rush to hospital emergency rooms, thinking that the intestine is prolapsing, that is, coming out.

TREATMENT

The basic idea is to kill or weaken the worm with medication and then flush it out with purges. The treatment is not so much painful as it is destructive of dignity. In order for the treatment to be successful, the scolex must be expelled. If the scolex remains, it will continue to produce proglottids and the treatment will have to be repeated. The only way to find out if the scolex has been removed is to look for it in the material expelled from the host's intestine.

The search for the small (one millimeter) scolex is, as far as the physician is concerned, the most difficult part of the treatment. Tapeworms subjected to anti-helminthic drugs do not come out in one neat piece; they emerge in several pieces, along with other intestinal contents that tend to obscure the scolex and make its finding a tedious task. In order to enhance the visibility of the scolex, as much of the intestinal content as possible is removed before giving the drugs, using techniques such as soap-water enemas and saltwater purges which, as far as the patient is concerned, are the most difficult and bothersome aspects of the whole procedure. However, clearing the intestine also serves to get more of the worm in contact with the drug.

A drug called niclosamide (Yomesan) is the one frequently used for most human tapeworm infections. Also used is quinacrine hydrochloride (Atabrine). Typically, the day before the treatment the patient is put on a liquid diet. The evening meal on the day before the treatment is usually eliminated. Some physicians order the soap-water enema at this time.

The drug is usually given in the morning. It might be given in split doses, some in the morning and some later. The drug makes some people nauseated. Splitting the dose cuts down the chances of vomiting. Two or three hours after the last dose, a saltwater purge is sometimes given in hopes of flushing out the dead or distressed worm. If the saltwater purge does not produce the scolex, another soap-water enema may be given. Sometimes the scolex and other worm parts get stuck in the large intestine on their way out.

If the scolex is found, the treatment is considered to be successful, but there is always the possibility that more than one worm is present. Happily for *T. saginata* hosts, one worm only is the usual situation. However, follow-up examinations are made for several weeks after the treatment, just in case there might have been more than one.

Often the dead worm is digested before it emerges, particularly if niclosamide is used. For this and other reasons, many physicians believe a search for the scolex is unnecessary. Follow-up examinations will reveal the success of the treatment.

FACTS AND FIGURES

Beef tapeworm is one of the more frequently seen tapeworm infections in the United States. However, it is not considered to be a major problem. Many, if not most, American physicians never knowingly see a tapeworm infection in their entire careers. Nevertheless, beef tapeworm is by no means extinct in North America.

Beef tapeworm is perpetuated by people's preference for rare beef. And although most beef is safe, there is always the possibility that any beef might contain some cysticerci.

Cysticerci are visible if you look for them, and beef heavily infested with them has a characteristic appearance that beef inspectors call "measly beef." Most beef sold in the United States is federally inspected and a search for cysticerci is part of the inspection routine. The majority of *T. saginata* infections come from eating home-grown beef, particularly grass-fed cows. Cows are most likely to ingest cysticerci if they eat grass that has been contaminated by human wastes. Even though most Americans use toilets, many cities are still guilty of dumping raw sewage into rivers. Rivers overflow from time to time and, when a sewage-laden river overflows, a film of sewage that might harbor a variety of worm eggs is left on the riverbanks in the area that was flooded. Cows grazing in areas that have been covered by polluted rivers can ingest tapeworm eggs along with other infective material left there by the receding waters. If it were not for the fact that grass-fed beef is tough and stringy, the incidence of beef tapeworm might be much higher than it is.

Beef tapeworm is not limited to out-of-the-way rural areas. The school secretary lived in an affluent suburb of a major northeastern city.

There are a number of ways to prevent beef tapeworm infections. Cysticerci are usually killed if the beef containing them is kept at $-10°$ C. ($14°$ F.) for five days. Thorough cooking, as mentioned earlier, also kills the cysticerci. Pickling the meat in a 25-percent salt solution also kills them.

The best and cheapest way to prevent tapeworm is to cook beef until all the red is gone. The rare-beef habit is firmly established in the United States, however, and concern about tapeworm is not likely to stop it. This state of affairs does not worry too many public health specialists. The chances of getting tapeworm from inspected beef are low and, for most hosts, harboring a *T. saginata* or two is usually not a matter of great consequence. However, there is a remote possibility that a human being can be the "intermediate host"; that is, the cysticerci can become established in human tissues. This can be a dangerous situation, especially if the cysts are in the brain. Fortunately, beef tapeworm cysticerci in human hosts is a very rare condition, and one does not get it from eating beef, however rare, but from accidentally ingesting the eggs. Human cysticercal infections are far more common from the next tapeworm we will consider.

Taenia Solium

This tapeworm is similar to the beef tapeworm in many ways. One obvious difference is that the intermediate hosts are pigs rather than cows. The proglottids look very much like those of *T. saginata.* They are not as active as the beef tapeworm sections. A major difference is seen in the scolex, and this difference is useful in making identification. The scolex of *T. solium* is armed with a vicious-looking ring of hooks. These hooks help to secure the worm to the intestinal wall.

The life cycle is practically identical with that of *T. saginata*, except that the intermediate host is a pig. While humans are the only usual primary host (some apes and monkeys can be experimentally infected), other animals besides pigs can be intermediate hosts. These include dogs, sheep, deer, monkeys, cats, and rats.

When undercooked pork containing cysticerci is eaten, the scolex in the cyst evaginates—that is, turns right side out—and the production of proglottids begins. *T. solium* is usually about 2 to 4 meters (about 6 to 13 feet) long, and contains 800 to 1,000 proglottids. The gravid proglottids break off and pass out in groups of five or six. They seldom, if ever, emerge on their own power. Each proglottid can contain as many as 50,000 eggs. The gravid proglottids burst to release the eggs. Sometimes they burst before they emerge. The life cycle continues when a pig or other suitable intermediate host ingests the eggs.

SYMPTOMS

T. solium has been known to cause its host more discomfort than does *T. saginata.* However, as is the case with *T. saginata*, many pork tapeworm hosts never suffer any distress. The hooks on the *T. solium* scolex account for some of the problem. They can be irritating to the intestinal wall, and if there is more than one worm, as can indeed be the case with *T. solium*, the effect can be compounded. Also, *T. solium* is thought to be a bit more "sandpapery" than *T.*

saginata, a circumstance that can result in irritation and abdominal discomfort. There have been occasional instances of a *T. solium* perforating the intestine.

TREATMENT

Treatment is much the same as for *T. saginata*. It is extremely important, however, that vomiting be prevented. Vomiting could possibly result in regurgitation of eggs and subsequent swallowing of them by the patient. Swallowing of *T. solium* eggs is quite likely to lead to the establishment of cysticerci in the tissues. Anti-nausea drugs are usually given one hour before treatment.

FACTS AND FIGURES

T. solium is a rare tapeworm in the United States. It is fairly common in some parts of Mexico, Central and South America, and Africa. The same prevention measures that apply to *T. saginata* can also be applied to *T. solium*. Pork tapeworm cysts, however, are a hardier breed. They can survive freezing at 0° C. to −2° C. (32° F. to 28° F.) for months. To kill them, the pork must be frozen at −10° C. (14° F.) for at least four days. Cooking the pork thoroughly, until *all* the pink is gone, also kills the cysticerci. It is possible to get tapeworm and trichina from the same pork.

T. solium Cysticercosis

The main danger from *T. solium*, as far as injury to health is concerned, is not the adult worm but the cysticerci. Infection with cysticerci—cysticercosis—is the result of ingesting the eggs of the tapeworm. The eggs can come from food or water contaminated with human sewage. A host can also swallow the eggs of his or her own *T. solium.* Sometimes the eggs can be forced into the stomach through vomiting or a reversing of intestinal movement (reverse peristalsis). There is also the possibility of infection from contact with *T. solium* hosts with careless and unclean habits. That careless or unclean host can self-infect. If a *T. solium* host does not thoroughly wash his or her hands after going to the bathroom, the possible consequences are obvious.

Once the eggs are ingested, the larval worms bore into the stomach wall and get into the blood circulation. Cysticerci then form in the tissues. The situation is usually not too serious if the cysts become established in the muscles. There might be a few muscle pains, some weakness, dizziness, and weight loss, but that's the extent of it. Eventually, the cysts become calcified and the larval worms therein die. The real trouble comes when the cysticerci form in and on the brain. Ironically, the cysts cause little trouble when they are alive. Pressure and swelling might occur, but this does not usually bother the patient. The problems begin when the worms die, and this may not occur until five to eight years after the cysts form. Some cysts, however, have lived as long as twenty years.

The symptoms are the result of reactions to the toxic products formed by the dead cysts. The patient can have seizures that are much like those of epilepsy. The body might go into spasms and the patient lose consciousness. Depending on where the cysts are located, the patient might have personality changes, insomnia, irritability, loss of intelligence, headaches, poor vision, and other problems. Sometimes a cyst forms in one or both eyes. Eye cysts do not necessarily result in loss of vision, but it can occur. Abnormalities of vision are more common than blindness.

TREATMENT

The only treatment available is surgery. Surgical removal of cysts in the eye is most feasible. When there are many cysts in the brain, surgery is not particularly useful. Cysticercosis is prevented by preventing adult *T. solium* infections. When *T. solium* is diagnosed in a patient, treatment should be prompt and complete.

Dibothriocephalus latus (or *Diphyllobothrium latum*)

The fish tapeworm is another example of a parasite that was carried from one part of the world to another by human movement. No slaves, however, carried this one around. It was brought from Europe to America by immigrants. Suitable intermediate hosts existed in the ponds and lakes of America and soon this large tapeworm was well established in the land of the free, where it still thrives today. *D. latus* is a good example of a parasite that has more than one intermediate host.

LIFE CYCLE

The adult lives in the small intestine of the host, usually in the jejunum and the lower part of the small intestine called the ileum. This worm is not as particular about hosts as are *T. saginata* and *T. solium.* In addition to human hosts, it does quite well in bears, dogs, cats, and several other mammals, including foxes, seals, and many other mammals that eat fish on a regular or occasional basis.

D. latus is the largest adult tapeworm that infects man. It is as much a monster as anything you might see flickering across a TV or drive-in movie screen. Composed of more than 3,000 proglottids, *D. latus* is usually about 3 to 10 meters (about 10 to about 33 feet) long. However, specimens as long as 15 meters (50 feet) and longer have been recorded. The gravid proglottids are wider than they are long, hence the name "latus," which means broad. The proglottids can be from 10 to 20 millimeters (almost an inch) wide. The scolex is somewhat spoon-shaped with two deep grooves that help it to hold fast to the intestine.

The eggs are "laid" by the gravid proglottids while they are still in the intestine. About a million eggs a day can be produced by a large *D. latus.* The proglottids disintegrate after shedding the eggs. Sometimes empty and partially empty proglottids, or pieces of proglottids, pass out in the feces.

The eggs must be in water in order to hatch. A swimming larval form, called the *coracidium*, hatches from the egg. If ingested within one or two days by tiny freshwater animals, such as cyclops or diaptomus, the cycle continues. The coracidium moves into the body cavity of the tiny cyclops, where it develops into another larval form called a *procercoid.*

The next stage in the life cycle occurs when the cyclops or diaptomus is eaten by any number of species of freshwater fish, including salmon, pike, pickerel, whitefish, turbot, and trout. When the fish digest the little animal, the procercoid is set free to penetrate the intestinal wall of the fish and to become established as another larval form in the muscles of the fish. The larval stage in the fish is the *plerocercoid* stage.

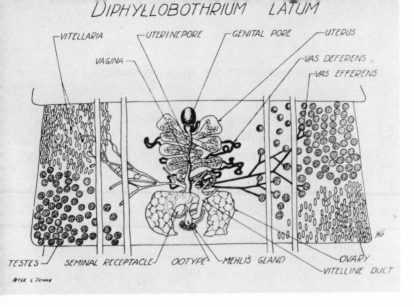

DIPHYLLOBOTHRIUM LATUM

VITELLARIA — UTERINEPORE — GENITAL PORE — UTERUS

VAGINA — VAS DEFERENS

VAS EFFERENS

TESTES — SEMINAL RECEPTACLE — OOTYPE — MEHLIS GLAND — OVARY

VITELLINE DUCT

AFTER L. DEMME

*Diagram of a mature fish tapeworm proglottid. Although a tape-
worm is often thought of as a simple organism, its internal structure
can be quite complex. Practically every structure in a mature tapeworm section functions in reproduction. Note that the section contains both male and female reproductive organs.*

THE CENTER FOR DISEASE CONTROL

The plerocercoids, commonly referred to as *spargana* (the
singular form is *sparganum*), are whitish, twisty organisms
about 15 millimeters (about five-eighths of an inch) long and
3 millimeters thick. Since freshwater fish tend to eat quite a
few cyclops and diaptomus, a single fish can be literally
loaded with plerocercoids. The plerocercoids can be seen if
an infected fish is held up to a strong light. If the fish is eaten
by a suitable host, such as a human being, and the plerocercoids are not killed by cooking, they will develop into adult
tapeworms in the intestine.

Plerocercoids can survive even if the fish is eaten by another fish. Smaller fish are generally eaten by large fish, and the plerocercoids are passed along from one fish to another. The plerocercoids do not go through any changes in this fish-to-fish transit.

Once in the intestine of the primary host, *D. latus* grows rapidly. Thirty proglottids produced in a day is not uncommon. Eggs will start to appear in the stool of the host about three to five weeks after the plerocercoids are swallowed.

Most *D. latus* infections involve one worm. However, multiple worm infections are not uncommon. Fortunately, the more worms there are, the smaller they are likely to be.

SYMPTOMS

There are probably thousands of *D. latus* hosts who go through life unaware that they were ever home and hearth to forty or more feet of tapeworm. On the other hand, some hosts have symptoms ranging from the merely annoying to the quite severe. As is the case with beef and pork tapeworms, many people start to have symptoms only after they know the tapeworm is residing in their very own interiors. These symptoms include digestive disorders, nervousness, nausea, and vague abdominal discomfort. Some people have more definite symptoms, such as diarrhea, malnutrition, and weight loss. Occasionally, there can be so many worms that the intestine is blocked.

D. latus has a particular affinity for vitamin B_{12}. This vitamin is needed for the development of red blood corpuscles. A deficiency of the vitamin can result in a very severe anemia, and that is just the problem many *D. latus* hosts face. Fish tapeworms can absorb from 80 to 100 percent of the vitamin B_{12} the host ingests in his or her food. The result is a type of anemia similar to pernicious anemia, a severe, potentially fatal form of the disease. People who have pernicious anemia can live normal lives by taking vitamin B_{12}. For *D. latus* hosts, the cure is to get rid of the tapeworm. Taking vitamin B_{12} pills will do little, if any, good; the tapeworm will just continue to soak up the vitamin. Injections of the vitamin, however, can be of benefit.

The amount of vitamin B_{12} the worm takes in is related to where the worm is located. If the worm is in the jejunum, no more than 145 centimeters (about 5 feet) from the mouth, the amount of vitamin B_{12} the worm steals from the host is considerable. However, if the worm is forced to move downward because of treatment or some other reason, the amount of vitamin B_{12} it takes in is considerably reduced and the anemia is accordingly relieved. The incidence of this kind of anemia in *D. latus* hosts is low in North America. It is highest in Finland.

TREATMENT

The treatment is similar to that used for the *Taenias*. Again, the critical factor is expelling the scolex. Since multiple worms are more common with *D. latus* than with *T. saginata*, the degree of success cannot be known until the stool is checked for eggs several weeks after the treatment.

FACTS AND FIGURES

D. latus is found in many parts of the world. In Europe it is particularly prevalent in the countries on the Baltic Sea. It is also found in the lake country of Switzerland, and in other countries that border on the lakes, as well as in areas along the Danube River. In Asia it occurs in various parts of the Soviet Union, Manchuria, Japan, and the Near East. The tapeworm was not known to occur in North America until late in the nineteenth century. Cases have been concentrated in the Great Lakes region, particularly Michigan and Minnesota. Other states in which it has been concentrated include Alaska, California, and Florida. However, cases have been reported in almost all the states.

It is believed that *D. latus* was brought to the United States by immigrants from Baltic-area countries, particularly Finland. Many Finns were infected from the local custom of eating smoked raw freshwater fish. The practice of dumping raw sewage directly into lakes soon had *D. latus* well established. Many lakes in Canada also became filled with heavily infected fish.

A curious concentration of *D. latus* infection existed in the United States among Jewish women who were recent immigrants from Europe. A pattern often seen in families was heavy infections in the mother and older daughters, but not in the males of the family. The reason for this pattern of infection was the method of preparing gefilte fish, an ethnic Jewish food. Gefilte fish is prepared by chopping the fish finely, adding spices, and shaping the chopped fish into cakes. The fish cakes are then boiled. Common practice was to taste the fish in the course of cooking to see how it was coming along, to see if additional spices were needed, and so on. Daughters, helping their mothers in the kitchen, would also take a turn at tasting the fish in preparation. The result of all this taste-testing was repeated reinfection. Gefilte fish is a traditional Sabbath dish, a circumstance that exposed the women to reinfection every week. Infections among Jewish women who followed traditional customs were extremely heavy. The men in the family were safe from *D. latus.* By the time they sat down at the table, the fish was cooked and any spargana that might have been in the fish were killed.

Today the incidence of infection among Jewish women is little (if any) higher than that of the general population. The reason for this drop is not so much any greater knowledge of the dangers of eating undercooked fish as it is changes in cultural patterns. Fewer Jewish women bother to prepare gefilte fish today. The granddaughters and great-grand-daughters of immigrant Jewish women who still serve gefilte fish are more likely to serve prepared gefilte fish picked up from a supermarket shelf, rather than spend hours making it at home, sampling as they work.

Fish containing plerocercoids is safe to eat if the fish is cooked for at least ten minutes, either fried, boiled, broiled, or baked. Raw fish dishes such as the Japanese sashimi and sushi should be limited to saltwater fish. Including freshwater fish in preparations of this kind is asking for some uninvited internal company.

Although fish tapeworm is not the kind of thing today's physicians are likely to come across every week, the worm is still with us and is not likely to vanish into extinction. Mini-epidemics of *D. latus* infections have occurred and can occur again. If economic conditions should ever force towns and municipalities to start dumping raw sewage into bodies of fresh water, the resulting epidemics might not be so small. Even in areas where sewage is treated, the life cycle is kept going in wild animal reservoirs. The infection could make the leap from these reservoirs into human intestines should raw fish dishes ever become popular.

Sparganosis

There are other species of *Dibothriocephalus* that can, in certain circumstances, exist in humans in the larval or sparganum stage. This condition is called sparganosis. Although rare in North America, sparganosis has been known to occur here, particularly in Florida. The species of *Dibothriocephalus* that causes sparganosis cannot grow to adult stage in man.

Sparganosis can be contracted from drinking water containing infected cyclops and similar organisms. The infection can also be obtained in some rather exotic ways. In some parts of the world, including the United States, treating wounds with a poultice prepared from freshly killed snakes or frogs was a common folk-medicine practice. If poultices of this kind are applied to open wounds, any worm larvae that might be in the snake or frog flesh can wiggle in through the open wound. The eating of raw snake flesh and tadpoles for some supposed medical benefit can also introduce the larvae. The larvae increase in size in the tissue and, in some cases, reproduce there by budding. They do not, however, become adult worms. They can cause extremely painful inflammation.

Sparganum proliferum is a form of the disease in which the spargana assume weird, fantastic, and grotesque shapes. This free-style monster larva can grow to considerable size. Pieces of it can break off and start to grow in other parts of the body. The effects can be similar to those of a proliferating cancerous growth.

Treatment of sparganosis is at present limited to surgical removal.

Hymenolepis nana

From the monstrous hugeness of the fish tapeworm we go to the smallest tapeworm for which man is commonly a host. *Hymenolepis* is usually about 20 millimeters (a little more than three-quarters of an inch) long. However, some specimens as long as 100 millimeters (about 4 inches) have been found. It is also the most short-lived of human tapeworms, having a life span of only a few weeks. What it lacks in size and longevity, it makes up in numbers. Infections of from 1,000 to 2,000 worms have been recorded. The usual number of proglottids is about 200.

LIFE CYCLE

Hymenolepis does not normally make use of an intermediate host. Its usual hosts are humans, mice, and rats. Some varieties of this worm can exist in intermediate stages in fleas and beetles of various species.

The worms live in the small intestine, in the ileum. Proglottids break up in the small intestine and the eggs are passed out in the feces. The eggs are infective as soon as they are released. If the eggs are swallowed by a suitable host, the first larval stage hatches from the egg and penetrates a *villus* (plural: *villi*). Villi are tiny, fingerlike projections in the small intestine. They serve to increase the area of absorption in the intestine. Inside the villus, the larval worm goes through another developmental period that lasts about four days. In that time it develops into another intermediate stage. That stage breaks out of the villus, becomes attached to the wall of the small intestine, and proceeds to grow into an adult tapeworm. The new worm is capable of producing eggs in about ten days to two weeks.

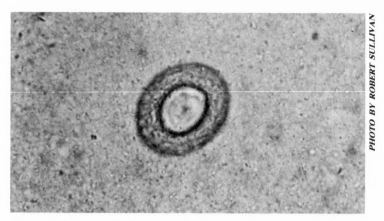

An egg of Hymenolepis nana, *the dwarf tapeworm*

Once *H. nana* gets started in a population, it is very difficult to get rid of it. The infection is direct, and hosts can keep reinfecting themselves by bringing eggs to the mouth on dirty hands. As is the case with many parasites, such as *Ascaris,* that have direct life cycles involving no intermediate hosts, children are among the chief victims. Sometimes children become infected with varieties more common in rats and mice. This kind of infection can come from playing in areas contaminated with rat droppings.

SYMPTOMS

As usual, a few cause no problems, but infections of a few dwarf worms are the exception rather than the rule. Heavy infections produce dizziness, diarrhea, loss of appetite, and nausea. In very heavy infections there can be internal reinfection. Eggs will penetrate villi in the host in which they were produced, rather than pass out in the feces.

TREATMENT

Expelling the worms is the only way to stop this cycle of reinfection. Niclosamide and paramycin are drugs that have been used to eliminate *H. nana.*

Personal cleanliness, training children to use toilets and to wash their hands afterward, and rat control are some ways to control *H. nana.* However, when there is internal reinfection, these measures don't accomplish anything. As is the case with *Strongyloides* the rate of infection among severely retarded, institutionalized people is quite high.

FACTS AND FIGURES

The World Health Organization estimate of *H. nana* infections is about 20 million. It is found everywhere in the world.

Echinococcus granulosus

This parasite presents somewhat of an organizational problem to the author. It could just as well be included in Chapter 9, which is a discussion of parasites people can obtain from associating too intimately with pet animals. *Echinococcus* is certainly one of those. However, there are some arguments for including it here. *Echinococcus* is a much more serious and widespread health problem in many parts of the world than most of the occasional infections discussed in Chapter 9. Also, man is seldom, if ever, the primary host for the adult form of this tapeworm. The problems arise when man is the intermediate host. It definitely merits consideration as one of the major human tapeworm infections.

LIFE CYCLE

The primary hosts of *E. granulosus* are dogs, wolves, foxes, coyotes, and an occasional cat. The adult worms, which live in the intestine of the primary host, are among the smallest known tapeworms of mammals. The adult worm is about 6 millimeters long (about one-quarter inch). It is usually composed of three segments, including the head and scolex. One segment is generally immature, one mature, and one gravid. They can exist by the thousands in one dog, but they seldom seem to be of concern to their canine host. The eggs pass out of the host in the feces. The nature of the hosts practically assures that the eggs will be deposited on the ground.

In nature the most frequent intermediate hosts are sheep. Other animals that can be intermediate hosts include other herbivores, such as cattle and horses, and humans.

There is a significant difference between the *E. granulosus* intermediate host stages and those of most other tapeworms. Although most intermediate host stages, such as *Taenia* cysticerci, are inactive, *E. granulosus* intermediate stages actively grow and reproduce in the intermediate host, with potentially disastrous results.

When the intermediate host swallows the eggs, the larvae hatch, penetrate the wall of the intestine, and are carried in the bloodstream to all parts of the body. The tiny (around 0.035 milliliter) larva then goes through a slow but remarkable, and potentially deadly, change. That tiny little egg can turn into a monster of horrible proportions. In about five months after ingestion, it grows into a roundish structure about 10 millimeters in diameter (one-half inch). At this stage and beyond, the structure, which is the larval stage of the tapeworm, is called a *hydatid cyst.*

If left undisturbed, the cyst will continue to grow. The cyst is covered with a wall made of a thick, tough material. The thick material covers a thin layer of cells called the germinal epithelium. This germinal epithelium is an actively growing layer of cells. From it, masses of cells grow into the cyst. These masses of cells grow in globular shapes called *brood capsules.* Thousands of scolices (the plural of scolex) form from the wall of the brood capsules. Sometimes additional cysts, called daughter cysts, form freely inside the cyst. These daughter cysts produce brood capsules of their own and can produce more daughter cysts, which could be called granddaughter cysts.

The entire cyst is filled with a fluid. The contents of the brood capsules are thousands and thousands of scolices, which are called hydatid sand.

Over the years the cyst becomes quite a complex, filled with brood capsules, daughter cysts within daughter cysts, each with its own brood capsules and scolices or hydatid sand. The fluid within the cyst exerts an enormous amount of pressure on the cyst walls. Sometimes the cyst bursts, releasing hydatid sand, brood capsules, daughter cysts, and pieces thereof into the host's body. This material can be carried to other parts of the body and develop into more cysts. Sometimes, however, the cysts are sterile. They stay small and produce no brood capsules.

The life cycle is continued when the primary host ingests scolices—that is, hydatid sand—in the course of eating an animal harboring the cysts. That turn of events is not so farfetched as it might seem. Many of the natural intermediate hosts of *E. granulosus*, such as sheep, are prey for wolves, foxes, and coyotes. Predators generally eat their kill's internal organs first. Internal organs, such as the liver, are fre-

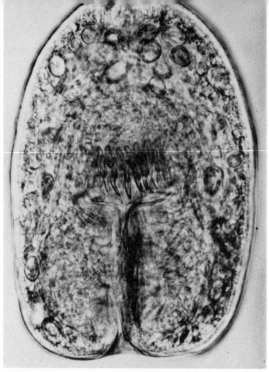

A highly magnified photomicrograph of a single scolex from a hydatid cyst, the intermediate host stage of Echinococcus granulosus. *These scolices are a major component of hydatid sand, the material inside the cyst. A single cyst can contain hundreds of thousands of scolices. The scolex hooks can be seen in the center.*

THE CENTER FOR DISEASE CONTROL

quent sites of hydatid cysts. Since one hydatid cyst can obtain 2 million or more scolices, the levels of infection in primary hosts can be heavy indeed.

Dogs might sometimes kill and eat a sheep, but predation is not the primary source of infection for dogs. Most dogs get *Echinococcus* from eating offal—internal organs of animals that have been slaughtered for meat.

Dogs are the main source of human hydatid infections. Where there are dogs there are dog droppings, and where there are dogs with *Echinococcus* there are tapeworm eggs in those droppings. Some dogs eat their own and other dogs' fecal matter and, in so doing, can get a mouthful of eggs. People will let dogs lick them and eggs can thus find their way into human interiors, either by direct dog licks on the mouth or by licked hands carrying the eggs to the mouth. Even so friendly and innocent an act as petting a dog can result in egg-contaminated hands. Dogs like to roll in the grass and, since the whole wide world is a dog's bathroom, all manner of infective material can be picked up from a dog's fur. Of course, getting a hydatid cyst in this way is most likely to occur in those areas where *Echinococcus* is prevalent but, no matter where you are, a good washing of the hands after contact with dogs is never a bad idea.

Of course, hydatid cysts growing in human hosts represent a dead end for the parasite. Instances of dogs eating people are extremely rare indeed and, contrary to what you have heard in fairy tale and fable, people are not a major food of wolves. Real, authenticated cases of wolves attacking and eating people are just as rare, if not more so, as cases of dogs attacking and eating people. Instances of dogs and wolves eating human flesh have occurred during wars and other catastrophes when there were likely to be bodies lying about. Usually tales of wolf packs attacking travelers in the forest are just that—tales.

The consuming of human flesh is not a factor in the life cycle of *E. granulosus.* However, the presence of hydatid cysts in human bodies is a matter of concern to those who harbor them. Serious, often fatal disease is the frequent result.

SYMPTOMS

The seriousness of hydatid disease is a matter of where the cysts are located, how many there are, and the size of them. Hydatid cysts can occur anywhere but, as mentioned before, are most likely to grow in internal organs such as the liver, spleen, and lungs. They can, however, grow practically anywhere in the body. They have been found in the brain, in bone, in the eyes, the heart, the body cavity, kidneys, and elsewhere.

There is no average size for a hydatid cyst. They can occur in many sizes and shapes, depending upon where they are located. Cysts with diameters of 20 to 30 centimeters (8 to 12 inches) are not uncommon. Some as large as basketballs have been removed from human hosts. However, many years, perhaps forty or fifty, are required for hydatid cysts to get that big. Most of them seem to be about the size of an orange.

The natural tendency of hydatid cysts is to grow in a spherical shape, but they can grow in many shapes and directions if body structures in the host restrict their growth in one or more directions. Those that grow in a free-form way are called *alveolar cysts.* A cyst of this kind has many branches and lobes. Pieces of it can break off and grow into new cysts that can also grow in practically any direction. Some parasitologists are of the opinion that alveolar cysts are not intermediate hosts of *Echinococcus granulosus* at all, but are those of a related species called *Echinococcus multilocularis.*

Hydatid cysts that grow in bone follow the path of least resistance and grow in the general shape of the marrow cavity inside the bone. Sometimes the cysts send gelatinous projections into the body of the bone.

Small hydatid cysts, depending on where they are, usually cause no symptoms. Small cysts in the liver can go unnoticed for the lifetime of the host and be discovered only on autopsy. However, if the cysts, even small ones, press against the bile ducts, all sorts of symptoms resembling gallbladder trouble could occur. Bile could back up into the body, causing the yellowish cast of jaundice. Pain could come from a gallbladder distended with bile. Ordinarily, the gallbladder contracts from time to time to squirt the bile into the small intestine, where it helps to digest fats. Cysts in the lungs can, if they are large enough, cause shortness of breath, pain, and excessive coughing that brings up blood. Cysts can also press against major blood vessels, cutting off or reducing the blood supply to an organ.

Serious problems can occur if the cysts grow in the brain and other parts of the nervous system. Symptoms include seizures like those of epilepsy. There might be partial paralysis, difficulty in speaking, involuntary shaking movements, and other nervous system problems.

The rupture of a cyst can bring on many complications. The liquid, hydatid sand, and other material bursting forth from the cyst scatters into the bloodstream and all parts of the body. Allergic reactions to all this material can be very severe. Sometimes the reaction is so severe that the patient goes into what is called anaphylactic shock, which can be followed by sudden death. Ruptured cysts can also lead to bacterial infections. Cysts can rupture from their own pressure, a blow from the outside, careless surgery, coughing attacks, or almost anything that might disturb the cyst.

TREATMENT

Hydatid cysts are frequently found by accident. Sometimes they show up on X rays taken during routine medical examinations. Symptoms produced by large cysts might also result in an examination that includes X rays and other diagnostic procedures, such as scans and ultrasound examinations. All of these procedures can reveal the presence of the cyst, but do not positively identify it as a hydatid cyst.

The only way to treat a hydatid cyst is to remove it surgically. Often a growth that shows up on the X ray is not identified as a hydatic cyst until after it is taken out. If the surgeon does not know that he or she is dealing with a hydatid cyst, the situation is potentially dangerous. Accidentally cutting into the cyst can cause it to burst, much like a balloon, liberating hydatid sand and other material capable of growing into more cysts into the body of the patient, or causing allergic reactions.

Methods of positively identifying an internal growth as a hydatid cyst are not totally reliable. Often, the diagnosis is a matter of logical deduction, putting together the pieces of a puzzle. For example, an X ray might reveal a growth in the liver. Blood tests for certain chemicals often present in the blood of cancer patients will be negative. The patient might come from an area where hydatid disease is prevalent, or might have a history of long association with dogs. All of these factors point to the possibility that the growth is a hydatid cyst.

A biopsy, the inserting of a long needle into the growth and aspirating out some contents, is a frequently used diagnostic method. However, inserting a needle into a hydatid cyst could be dangerous; it just might burst. If the biopsy is done carefully, the resulting diagnosis is highly reliable. Microscopic examination of the fluid withdrawn from the cyst will reveal hydatid sand.

There are various blood tests that are sometimes reliable as indicators of hydatid cysts. However, there are many other conditions that could give false positive results.

Surgical removal of the cyst is not always possible. Sometimes the cyst becomes very intimately associated with the surrounding tissue. Important blood vessels can be sinuously wrapped around the cyst or through the many branches of an alveolar cyst. A procedure sometimes followed is the injection of formalin into the cyst. The formalin kills the little scolices and the germinal membrane that produces more brood capsules. Doing so makes removal of the cyst safer, and renders the cyst inactive if removal is impossible. Another technique used for nonremovable cysts is the injection of antigens that neutralize the material inside. Often the cysts atrophy after this treatment.

FACTS AND FIGURES

Hydatid cysts in people are a kind of evolutionary accident. The *Echinococcus* species get nowhere when the cysts are locked up in the tissues of a human being. Nevertheless, this accident occurs with alarming frequency in some parts of the world. Just how many people have cysts is not known; many are never diagnosed. Hydatid disease occurs mostly in the sheep-growing regions of Australia, New Zealand, and the Middle East. The association between man and dog is particularly intimate among shepherds. The practice of feeding sheep offal from slaughtered sheep to dogs serves to perpetuate hydatid disease quite nicely. Hydatid disease also occurs in South America. The disease is by no means absent from the United States and Canada. It occurs in sheep-growing regions such as Utah and exists in immigrants from sheep-raising areas.

The most obvious way to prevent hydatid disease is to avoid intimate contact with dogs. Such avoidance is not always easy. Dogs express affection by licking, and most people are appreciative of a dog's affection. The chances of picking up *E. granulosus* (not to mention a host of other parasites discussed in Chapter 9) from a dog's attentions are reduced by keeping your hands away from your mouth during and immediately after a romp with Fido. Eating during such periods of dog-human closeness is definitely asking for trouble. Always wash your hands after playing with a dog, especially if you are going to eat with those hands.

Most sheep ranchers are aware of the hydatid problem, and many take preventive action. Not feeding sheep offal to dogs, or cooking it before using it for dog food, is an effective preventative. All dogs should be routinely wormed at least once a year. Twice a year is the better choice of frequency in sheep country. Iceland completely eliminated hydatid disease by passing laws requiring regular worming of dogs.

The evolutionary accident of hydatid cysts in human beings is certainly no fault of the worm's, but is due, rather, to man.

6.

Flukes

Fluke is one of those words that has a variety of meanings. Years ago it was used in pool halls to describe a lucky, wild shot, one that should have brought disaster to the pool player's game but instead brought victory. The word now means any strange, bizarre happening that produces results contrary to what could ordinarily be expected.

Fluke is also used to describe the digging parts of an anchor, the parts of a whale's tail, and certain species of flat fishes. In that context, its meaning is more like the old Anglo-Saxon word from which it came—*floc*, a word that referred to something flat.

When one learns that fluke means "flat," it is not hard to understand why it is used to describe a class of parasitic flatworms. You will recall that the flatworm phylum is called Platyhelminthes. In the last chapter, we looked at the cestodes; in this chapter we will deal with the trematodes—the flukes, many of which exhibit some marvelously strange and bizarre patterns of behavior. As is the case with cestodes, all trematodes are parasitic.

There is a class of free-living flatworms, the Turbellaria.

The best-known member of this class is the planarian. This small (20 millimeters long, almost an inch) black flatworm is familiar to most people who have had a high-school or college biology course. Somewhat arrow-shaped with two light-sensitive eyespots that give it a comical cross-eyed appearance, it is a favorite object of study in biology classes.

Flatworms are low on the evolutionary scale. They were probably among the earliest forms of life to evolve. There is much about them that is primitive. They do not have well-developed systems of organs. (You will recall that tapeworms have no digestive system.) Their digestive tubes have only one opening. Many of them have a primitive but reasonably effective nervous system, including some specialized structures such as the light-sensitive eyespots of the planarian. Primitive though they may be, many of the free-living forms are brightly colored and quite spectacular to behold.

Flukes were probably among the first parasites. A subclass of parasitic flukes that exists today is strongly suggestive of this pioneering role. These are the flukes in the subclass Monogenea. The name alludes to the simplicity of the life cycle. Reproduction is direct from one generation to another. There are no stages in intermediate hosts. The Monogenea are mostly parasites of fish and amphibians. They are commonly ectoparasites, although some live on the gills and some are found in the *cloaca*—an organ in fish that is somewhat like the large intestine of mammals. Others live in the urinary bladder of frogs and some turtles.

The Monogenea retain many of the characteristics of free-living animals. Many have eyespots and, in general, are much more complex in structure than tapeworms. The Monogenea do have some interesting structures that aid them in their way of life. The sucking disks and hooks used to hang on to their mobile hosts are marvelously varied in structure. As parasites go, their rate of reproduction is slow. They produce few eggs, which develop directly into forms capable of attaching on to another fish.

The subclass Digenea includes some of the most important parasites in the world, because many digenetic trematodes (i.e., Digenea) cause some of the most devastating diseases of humankind and domestic animals. The economic development of entire nations and regions of the world has been retarded by these parasites.

From the strictly biological point of view the digenetic flukes exhibit some pretty fantastic behavior patterns. Remember that these are not intelligent animals, and what they do is strictly instinctive, the product of millions of years of evolution. The patterns into which they have evolved, the ways in which they have become adapted to take maximum advantage of their host's life-style, ensures they will be with us for a long time—probably as long as there is a planet on which to live.

The word digenetic implies a complex life cycle involving complicated stages in intermediate hosts. That is the case with digenetic trematodes. Although there is no typical life cycle, there are similarities in the life cycles of all of them.

The adult worms live just about every place there is to live in practically every vertebrate animal. The eggs escape from the host in a variety of ways, and, with few exceptions, must reach water if they are to hatch. The eggs hatch into a free-swimming larval form called a *miracidium* (plural: miracidia). These swim around, and if they encounter suitable intermediate hosts (almost always snails) the miracidia enter them through direct penetration. Inside the snail the miracidium changes into a saclike structure called the *sporocyst*. Inside the sporocyst, more larval forms are produced through budding, an asexual form of reproduction. The sporocyst may produce more sporocysts or other forms called *rediae* (singular: redia). The sporocysts on rediae produce more rediae or another kind of infective larva called the *cercaria*. Cercariae enter the main host either by boring directly into the skin or assuming forms that are swallowed by the host.

This basic life cycle differs in details from one species to another, but the same remarkable thing is accomplished no matter what the details are. All the kinds of intermediate forms in the snail come from *one* miracidium that hatched from *one* egg. What this means is that one egg has the potential to become hundreds of thousands of infective larvae, each one of which is capable of infecting a new primary host. That egg is but one of millions an adult fluke can produce in its lifetime. These numbers might help to explain why some 250 million people are estimated to have flukes of one kind or another.

One might wonder, after learning a few facts about the colossal reproductive powers of flukes, why only 250 million of some four billion people on this planet have flukes. The number of infections in humans and other animals might be much higher but for a quirk in the adaptations of flukes. Although many species of flukes have a reasonably wide choice of hosts in which to live their adult lives, that choice is considerably more restricted for the larval forms. A miracidium can grow in only a few species of snails. In some cases, only one species or one variety of a particular species will do. This has kept most kinds of flukes out of the United States, Canada, Europe, and a large part of the rest of the temperate world. Most flukes, particularly the highly destructive blood flukes, are tropical animals. There are, however, hundreds of thousands of human fluke hosts and ex-hosts living in the United States. Many of these hosts are ex-military personnel who picked up the parasites while serving in tropical areas; others are people who once lived in tropical areas. There was, and still is, some concern that some species of snails in the United States might be able to serve as intermediate hosts for some of these flukes. Thousands of experiments in which attempts were made to infect local species of snails with many different kinds of tropical flukes showed that the flukes just could not make it in American snails. There is still some fear that mutations might occur that will allow the flukes to establish themselves in local snails. Some species of snails, particularly one in Florida, have kept public health officials a little nervous since the late 1940's, when it was shown that some blood flukes could go through a limited development in them. Although there are few biologists who think that tropical flukes will ever be a big problem here, a little nervousness is understandable. Some of these parasites, notably the blood flukes, are among the most serious health problems in the world.

Schistosomes

During the last twenty-five years, most internal parasites of man have been declining in status as major health problems. The schistosomes, or blood flukes, however, have shown every sign of going against this trend. In many parts of the world there was an increase of schistosome infections in the 1970's.

There are hundreds of species of schistosomes, three of which are major human parasites. Schistosomes usually live inside blood vessels. There are separate male and female worms; most other flatworms are hermaphrodites. The length of these worms varies from 6.5 to 25 millimeters. The diameter is seldom more than 0.5 millimeter. The males are generally broader than the females, but females are a little longer. The male and female worms do not look at all alike. The male is flattish, but it hardly ever looks flat. Most of the time, its body is folded the long way so that a kind of groove is formed running the length of the worm. The female is roundish, slender, and lithe. When male and female schistosomes pair off, the female settles down in the male groove (called the *gynecophoric canal*), where the male clasps her in a connubial embrace. In most species of schistosomes, the pairs are monogamous, remaining blissfully joined in a lifetime of lovemaking. Among one species of human schistosomes, however, the pairings are less permanent and somewhat more casual in nature. The more or less permanent joining of male and female schistosomes has the effect of making the pairs a hermaphroditic organism.

LIFE CYCLE

The three human schistosome species are *Schistosoma mansoni*, *Schistosoma haematobium,* and *Schistosoma japonicum. S. mansoni* and *S. japonicum* live in small veins near the intestine, and *S. haematobium* lives in blood vessels farther down in the pelvic regions closer to the bladder.

The female lays her eggs in the tiny blood vessels, after

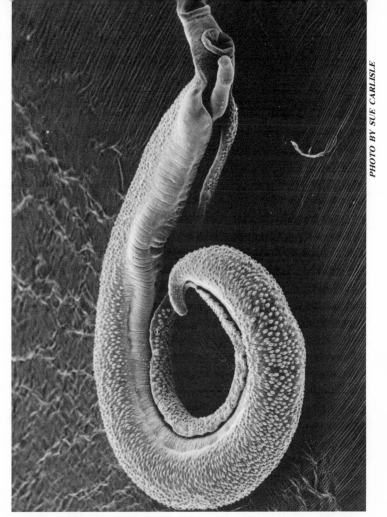

Scanning electron micrograph of a Schistosoma mansoni *pair. The male is the larger worm. The roundish, slender female is nestled in the "gynocophoric groove" of the male. The anterior or "head" end of the female can be seen poking out of the groove near the male's sucker disk. The mouth is above the sucker disk. The two worms can remain in this mating embrace for extended periods of time. However, unlike other species of schistosomes,* S. mansoni *pairings are not necessarily permanent.* **COURTESY DR. ANNA G. BENTLEY, UNIVERSITY OF PENNSYLVANIA SCHOOL OF MEDICINE.**

which the eggs work their way through the walls of the vessels. In the case of *S. mansoni* and *S. japonicum* the eggs end up in the intestine and are passed out in the feces. The eggs of *S. haematobium* break into the urinary bladder and go out into the world in the urine. There is some overlap in this escape route. Some eggs of *S. haematobium* get in the intestine, and a few *S. mansoni* and *S. japonicum* get into the bladder.

When the eggs reach water they hatch into miracidia, which have a limited time in which to find a suitable snail. Generally, their allotted time as free-swimming organisms is less than twenty-four hours. The suitable snail species are different for each species of schistosome. When the miracidia swim in the vicinity of the right snail, they make a high-speed dash for the snail. Like iron filings attracted to a magnet, they practically leap on to the snail. They burrow into the flesh, generally around the tentacles, much to their new host's discomfort.

Inside the snail, the miracidium changes into a sporocyst. The sporocyst produces daughter sporocysts that produce cercariae, the infective larvae. The formation of sporocysts and cercariae is accomplished through budding. As mentioned earlier, the reproductive potential is enormous. One miracidium, through the succeeding asexual stages in the snail, can give rise to as many as 200,000 cercariae in the lifetime of the host snail. All of the cercariae that spring from one particular miracidium are of the same sex.

The cercariae are about 0.275 millimeter long, including the anterior body section and a forked-tail section. The cercariae emerge in bursts from the snail. In some species, there is a periodicity; that is, the snails come out at certain times of the day. They enter the human host by boring directly into the skin. Of course, the human hosts must be in water in order to be exposed. People are infected while swimming, bathing, washing clothes, working in rice paddies (which must be flooded) and other irrigated crop fields, or wading across small bodies of fresh water. The cercariae can also enter in water the host drinks. In that case, the cercariae bore into the mouth and throat mucosa.

The cerariae bore in until they penetrate the skin, losing their tails as they do so. Eventually, they reach capillaries

Photomicrograph of a cercaria of Schistosoma mansoni. *Actual length of the cercaria is about 275 micrometers.*

THE CENTER FOR DISEASE CONTROL

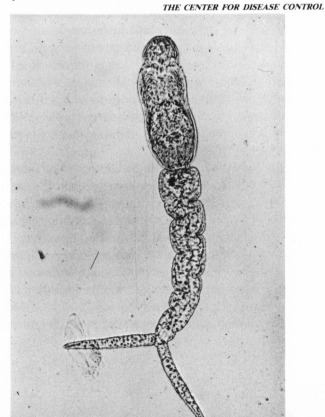

and are carried into the bloodstream, then in veins to the right side of the heart and into the lungs. Once inside the lungs, they wiggle into capillaries there and are eventually carried by the blood to the *portal circulation* via the liver. The portal circulation is made up of blood vessels that carry blood from the intestines to the liver. This blood is rich in absorbed food materials. The young worms stay in the rich, lush environment of the portal vessels for about three weeks, feeding and growing rapidly. They then swim "upstream"— that is, against the direction of blood flow until they reach their permanent homes, which, depending on the species, are small veins in and around the intestine or the bladder. There they feed on red blood cells. They are thought to live about thirty years.

I witnessed the drama of the schistosome life cycle on an almost daily basis during my senior year in college when I was a research assistant to Dr. Irving Kagan, then at the University of Pennsylvania. One of my jobs was to infect white mice with human schistosomes, *S. mansoni* and *S. haematobium.* The first step in this task was to retrieve cercariae-producing snails from the aquarium tanks stored in a basement room in the laboratory building.

The tanks looked just like any fish-hobbyist's aquarium tanks, but there was death and misery in those tanks. The snails could be seen plainly enough, placidly feeding on lettuce leaves. What could not be seen were the millions of infective cercariae swimming about in the water. Contact with that water could have resulted in getting schistosomiasis. Even a drop of water accidentally splashed on the skin might have been enough. Bottles of alcohol were nearby in the tank room; the alcohol was for applying to the skin in case of accidental exposure to the water.

The uniform for snail gathering was a long white coat and long rubber gloves bound at the opening with elastic bands. Although it was not required, I usually wore a head covering as well when I went into the tank room, leaving only my face exposed. Dr. Kagan had told me that if any cercariae began to bore into me I would feel some itching. Of course, the first time I retrieved snails, I started to itch unbearably as soon as I walked into the room, and I made an immediate dash for the alcohol. The itching was as much psychological as it was due to the sweating under all that protective clothing.

Snails were pulled from the water with a two-foot-long pair of forceps and placed in a covered container. The mice were infected by dipping their tails in water containing some of the snails. This was done by strapping the mice to a board and passing their tails through a hole in the board, so that when the boards were placed on top of the container of water, the tails were immersed. Later, the worms were retrieved from the mice and were used in tests for diagnosing human schistosomiasis.

Watching miracidia infect snails was particularly fascinating. I would put some snails in a small bowl of water, place the bowl under a low-powered microscope, squirt in a dropper full of miracidia, and watch. The miracidia would swim around rapidly and aimlessly. However, when they came within an inch or so of the snail, they would appear to go into a frenzy. They shot into the snail like bullets, and the snails twitched and writhed in very obvious discomfort. It was interesting to put the wrong type of snails in the bowl and watch how the miracidia ignored them. Just how an organism as simple as a miracidium can tell which snail is the right one is a fascinating question. Undoubtedly, the miracidia respond to particular chemicals given off by the snail.

These chemicals would be in the water near the snail, and when the miracidia come into contact with them, snail-boring behavior is activated.

SYMPTOMS

S. mansoni and *S. japonicum* Schistosomiasis is referred to by a number of terms in different parts of the world. In Egypt, it is generally called "bilharziasis." In the Far East, it might be called "Katayama disease." Disease caused by *S. haematobium* is called "visceral schistosomiasis" or "schistosomal haematuria." Everywhere it is called "snail fever," an allusion to the intermediate host.

The diseases caused by *S. mansoni* and *S. japonicum* are similar enough to be considered together. Schistosomiasis, caused by any of the three human species, is a devastating, debilitating disease. It does not kill its victims quickly and dramatically as do some bacterial and viral diseases, but instead slowly robs the victim of strength and vitality.

Troubles begin as soon as the worms enter. The prickling, burning sensation that accompanies their burrowing into the skin might be followed by skin irritations brought on by allergic reactions. Large numbers of them in the lungs can cause symptoms similar to pneumonia. While the worms are maturing in the intestine there may be symptoms such as diarrhea and fever.

The real trouble starts when the females start to lay eggs and the patient goes into the acute stage of the disease. The eggs break through the walls of the blood vessels and the walls of the intestine. During this activity, the patient suffers from fever, general "blahs," and pain in the abdomen. The area of the liver is very tender to the touch. There is also dysentery and diarrhea. If the worms lived in the intestinal lumen and the eggs passed out directly, they would not cause nearly as much trouble as they do. The irritations set up by the eggs penetrating through various tissues are the root of both the long- and short-range misery of schistosomiasis.

The disease then passes into what is called the chronic stage. Not all the eggs produced by the worms make it to the outside. If they did, the course of the disease would be much less severe. Many eggs are swept back in the portal blood-stream toward the liver. They lodge in various organs, particularly the liver. They also get stuck in the portal blood vessels. Growths called *granulomas* develop around the eggs. These growths eventually reduce the elasticity of the blood vessels, resulting in a condition similar to hardening of the arteries. The organs in which the eggs lodge are damaged and their functioning can be affected. The liver and spleen can enlarge. The degree of enlargement is usually slight in the acute stage, although, in some cases, it can be quite pronounced. The acute stage of *S. japonicum* is usually much more severe than that of *S. mansoni.*

The eggs that enter the intestinal wall and lumen of the intestine also cause their share of trouble. The tissue of the intestine can be permanently damaged, and the resulting inflammation and necrosis (death and decay of affected tissue) contributes to the general malaise and other symptoms of the disease. The toxic products given off by the worms also produce symptoms and complete the package of misery.

The chronic stage of *S. mansoni* can begin from eighteen months to twenty-five years after the acute stage, depending on how many worms are present. In areas of heavy infection, there is repeated reinfection of the worms, so the beginning of the chronic phase is generally closer to the eighteen months than the twenty-five years. A chronic disease is one that lingers; this is a term indeed descriptive of schistosomiasis that is not treated in the earlier stages.

S. *japonicum* produces about ten times as many eggs as S. *mansoni*, and the damage caused by S. *japonicum* is accordingly worse. In S. *japonicum* disease, the entire intestinal wall can be permanently damaged. The eggs of S. *japonicum* can really make a mess of the intestine.

During the chronic stage, there can be anemia and severe weight loss. At the same time, the spleen and the entire abdomen can swell to the point of grotesqueness. The victims characteristically have thin, emaciated arms, legs, and faces, and have an incongruously bloated appearance in the abdomen caused by the swelling. After a while, the liver actually shrinks in size, primarily due to blocked blood vessels, but the spleen often stays swollen. The victims of chronic schistosomiasis continue to have irritations of the intestine, diarrhea, and fever. The victim feels just plain lousy, and has little strength to work and make a living. Poor nutrition makes the disease worse. The unfortunate reality is that schistosomiasis is highly prevalent among poor people in poor countries, where much of the population suffers from chronic malnutrition.

S. *haematobium*   The eggs of S. *haematobium* leave the body by penetrating the wall of the bladder, entering the lumen of that organ, and exiting in the urine. Consequently, the major damage from this species of schistosome is to organs of the urinary and reproductive systems. Damage to other organs, such as the liver, may also occur, but this damage is usually not as severe as that produced by S. *man-*

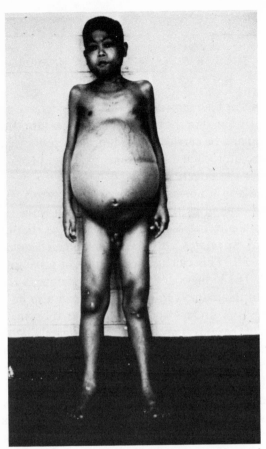

This boy, infected with Schistosoma japonicum, *shows the characteristics of the advanced stages of the disease. His liver is abcessed and grossly swollen.* **THE CENTER FOR DISEASE CONTROL**

soni and *S. japonicum*. However, *S. haematobium* can sometimes cause some rather severe damage to the lungs during the stage of the life cycle when it is moving through those organs.

S. haematobium adults live in the blood vessels called the pelvic veins. The eggs move through these veins to the urinary bladder. Over the years, the damage to the bladder and other parts of urinary and reproductive systems builds up gradually. The bladder wall may become ulcerated and there is frequently blood in the urine. Eggs can get into the general circulation and be carried to the lungs where they can cause more discomfort. These symptoms can begin ten to twelve weeks after the infection.

The walls of a normal, healthy bladder are very elastic. The walls expand as the bladder fills with urine, and it is blown up somewhat like a balloon. One of the long-term effects of *S. haematobium* disease is the calcification of the bladder walls. This calcification results in a painful loss of elasticity. The kidneys and the ureters, the tubes that carry urine from the kidneys to the bladder, can also be affected. Sometimes the ureters and urethra, the tube that carries urine out of the body, may become so damaged that they may be partially or completely blocked. Organs of the male and female reproductive system, such as the uterus in females and structures that produce sperm cells and semen in males, may also be damaged. Sometimes the penis and scrotum become swollen in a way that is suggestive of elephantiasis.

As the disease progresses, the victim may begin to notice blood in the urine. Urination becomes painful. There is frequently fever, a general feeling of listlessness, sweating, weakness, and a dull pain in the lower abdominal region. There is often bacterial infection in the damaged areas of the urinary system.

Although the overall effects of *S. haematobium* can be less severe than those of *S. mansoni* and *S. japonicum*, *S. haematobium* can lead to additional complications. There is evidence that *S. haematobium* increases the chances of cancer in the bladder and other parts of the urinary system. It is not clear whether this disposition of cancer in *S. haematobium* victims comes from direct damage to the cells caused by the worms, or from viruses the worms carry.

TREATMENT

Potassium antimony tartate (tartar emetic) is one of the few drugs that is effective against *S. mansoni* and *S. japonicum*. It is usually given by intravenous injection. Schistosomes are tough, and though tartar emetic usually has the punch to kill many of them, that punch can also do a job on the patient. The toxic side effects of tartar emetic can be worse than the disease. Vomiting, fainting, and severe coughing often occur, and sometimes the patient is prostrated—unable to move. There are drugs, such as Astiban, that are easier on the patient, but they are also easier on the schistosomes.

In the late 1960's and early 1970's, new and often dramatically effective treatment was developed. The worms are first immobilized by doses of tartar emetic. Then portal blood containing the worms is tapped off and sent through a filter. The blood is returned to the patient's body, but the filter traps the worms and keeps them out. Many worms have been removed with this method, but it doesn't always remove all the worms.

The search for new drugs lethal to the worms and less so to the patient continues. One such drug being tested during the 1970's was oxamniquine. In many cases it has been very effective against the worms and has produced few side effects.

FACTS AND FIGURES

In the archives of the state department is a report stating that, in 1950, the recently victorious (1949) Chinese Communists had massed hundreds of thousands of troops along the coast in preparation for an assault on Taiwan. According to the report, the attack was canceled because between 30,-000 and 50,000 troops were sick with acute schistosomiasis.

It may be years before that story can be completely authenticated, but if it did happen it would not have been the first time schistosomes interfered with war and other activities of humankind. An outbreak of schistosomiasis among American troops during the invasion of the Philippines in World War II severely hampered the 1944 effort to recapture these islands from the Japanese.

Schistosomiasis is very much with us today. The World Health Organization has designated it as one of the world's major health problems. *S. mansoni* is another parasite that was transported from Africa to the western hemisphere by the slave trade. It is found in West Africa (including the old slave coast and in a band, extending through Nigeria, southern Chad, and Sudan Republic, to the east coast, including Tanzania, Mozambique, western Zaire, and Madagascar). There is a particularly heavy area of infection in Egypt north of Cairo. It also occurs in Israel, Saudi Arabia, and Yemen. In the Western Hemisphere, it is very heavy in Venezuela and parts of Brazil. It is also found in Guyana, Surinam, French Guiana, Puerto Rico, Haiti, Dominican Republic, Jamaica, and some other Caribbean islands.

S. haematobium is found in Africa in much of the same area as *S. mansoni*, but its range is a bit more extensive than *S. mansoni*'s. There are concentrations in Libya and Morrocco. It is found all along the Nile Valley in extremely heavy concentrations, and there is a moderate area of concentration in eastern Angola. There is a very heavy infection area along the Euphrates River in Iraq, and a small intense area in India.

S. japonicum has a very limited area. It is found only in a few areas of eastern China, Japan, Burma, Malaysia, Indonesia, and the Philippines. It can also occur in Laos, Cambodia, and Vietnam. Although its range is small, the disease caused by *S. japonicum* is particularly severe. It is the worm that stopped the invasion of Taiwan and disrupted the American operations in the Philippines during World War II.

For centuries, Egypt has had a serious problem with schistosomes. Schistosome eggs were found in a two-thousand-year-old Egyptian mummy. Egypt is also one of the poorest countries in the world, and that situation is due largely to schistosomes. In some parts of Egypt, the rate of infection among agricultural workers is from 90 to 100 percent. The rate in most of the agricultural areas along the Nile is about 60 percent. Many of the victims are likely to be hosts to both *S. mansoni* and *S. haematobium.* People carrying that kind of load cannot be very efficient workers.

Attempts to grow more food in Egypt have led to more schistosomiasis. The Aswan High Dam was built to provide hydroelectric power and irrigation water for growing crops. Irrigation ditches provide ideal conditions for snails to grow and multiply. These irrigation ditches are quite likely to be polluted with human wastes containing schistosome eggs. People working in the irrigated fields are constantly in contact with the water, and the cycle of disease and misery is perpetuated. It quite often comes down to a choice of food or flukes.

In the rest of Africa, people are infected while fishing or tending irrigated crops. In the Far East, people are infected while working in rice paddies. Rice paddies are flooded and are often fertilized with human wastes. The governments of

Egypt and other high-incidence countries have made attempts at control, but control measures are expensive and only partially effective. Keeping human wastes out of bodies of fresh water where the snails live would be very effective if it could be done. However, many of the areas of heavy schistosome incidence are primitive, and toilets are just not a part of life there. People relieve themselves wherever they happen to be, and the relieving is frequently done in the nearest body of water. In these same areas, using human wastes for fertilizer is often an economic necessity. However, allowing the night soil to sit for several days—long enough for the worm eggs to die—cuts down on the spread of schistosomiasis and other worms through this medium.

Nets and traps are used for capturing snails. There are also chemicals, such as copper sulfate and others, that will kill snails. Cercariae are killed by chlorinating the water. These chemicals, however, are expensive, and treating every body of water where there are likely to be snails is a monumental, practically impossible task.

Mass treatment of hosts would reduce the source of reinfection, but treatment is expensive, and it is not possible to find and diagnose all schistosome hosts. Equipping agricultural workers with protective clothing, such as waterproof boots, would be very effective but, again, the expense makes that control method unfeasible.

The problems faced by Egypt were undoubtedly a major motivating force for President Anwar Sadat's efforts to make peace with Israel. His country, faced with enormous problems of schistosomiasis, food shortages, and poverty, could not afford to keep on fighting wars. Until the 1973 Yom Kippur War those conflicts resulted in disaster for Egypt, and the 1973 War might also have been disastrous for Egypt had it not been stopped. Most of the soldiers in the ranks of the Egyptian army come from rural areas, and many of them, no doubt, have schistosomiasis. The weakened condition of many of the soldiers with schistosome disease might have contributed to the poor showing of the Egyptian military in these wars. Cooperation with Israel might help Egypt with its schistosome problem. Israel also once had a serious schistosome problem, but today, the incidence of the disease in Israel has been greatly reduced.

Although schistosomiasis is almost entirely a Third World problem, it is present in people practically everywhere. Jet airplanes have made the world smaller, and people travel more. One can be transported from the primitive squalor of an Egyptian village to a modern Western city in a matter of hours. In New York City alone, there are perhaps fifteen thousand people who harbor schistosomes. As pointed out earlier, there is little, if any, danger that tropical parasites in Puerto Rican Americans or more recently arrived immigrants from Asia will ever become a problem here. It is, however, important that newly arrived immigrants be treated for these parasitic infections as soon as possible. One positive aspect is that most of the people who fled Vietnam after the collapse of the South Vietnamese government were middle-class, urban people, less likely to be harboring parasites than people from rural, agricultural areas.

Malaria (page 192) was once the most widespread disease in the world. Since World War II, an international effort has dramatically reduced the impact of that disease. Malaria has by no means been eradicated, and there has even been a resurgence of the disease in some areas. A massive effort, international in scope, would be required to eliminate schistosome disease as a major health problem. It could be done, but it would require billions of dollars and, more importantly, the involved nations would have to forego their political and ideological differences long enough to cooperate in an effort for their mutual benefit.

Schistosome Dermatitis: Swimmer's Itch

Although human schistosomes are not native to the United States and Canada, many species of blood flukes are quite numerous here. These are parasites of birds and mammals such as muskrats. The intermediate hosts are species of freshwater snails that are common inhabitants of lakes and ponds in North America.

The life cycle of these schistosomes in much the same as that of human schistosomes. Miracidia bore into snails and, after asexual development in the snails, cercariae emerge and swim around until they die or find a host in which to bore.

When people swim in lakes and ponds, they are exposed to the cercariae. These cercariae can penetrate human skin, but they don't get very far. Burrowing into human skin is definitely a mistake for these organisms, but, of course, they are incapable of knowing that.

Most of these cercariae are trapped in the skin and die there. A few might make it as far as the lungs before they perish. There can be, and often is, an uncomfortable, painful reaction to those organisms that have ventured where they definitely do not belong. At first, there is a prickly, itching sensation, quickly followed by the reddish, raised weals on the skin. These usually go down in about an hour or less, becoming smaller, only slightly raised patches that, nonetheless, can still be itchy. The patches can break out into pustules two or three days later. The itching remains. There can also be generalized swelling all over the body. The condition usually clears up in a week or two, unless there is more exposure to cercariae or infection brought on by excessive scratching.

The severity of the reaction varies from person to person, and is a matter of the heaviness of the infection and the degree of allergic response of the individual. In mild cases, there is just a little prickling. In severe cases, the victim might get very sick, pass into unconsciousness, and have difficulty breathing. Repeated infections tend to sensitize the person; the reaction gets worse with each succeeding infection.

There is a seawater version of swimmer's itch called "seabather's eruption" in some places and "clam-digger's itch" in others. It is caused by cercariae of schistosomes that infect shorebirds. Swimmers can cut down their chances of getting the itch by vigorously rubbing the body with a towel as soon as they leave the water. The cercariae remain in the film of water left on the swimmer. They penetrate as the water evaporates. The behavior of the cercariae explains why children are more frequent victims of swimmer's itch than are adults. Children run in and out of the water constantly, and often don't take the trouble to use a towel between dips.

Parasitologists are investigating the possibility that repeated exposure to the cercariae that cause swimmer's itch can make a person resistant to infection by human schistosomes. At the beginning of 1980, the evidence of this effect was only suggestive and by no means conclusive. However, the evidence suggests the intriguing possibility that these cercariae might someday be used to make an anti-schistosomiasis vaccine.

Other Flukes

For a while, during the American military involvement in Southeast Asia, a great deal of attention was given to survival training. Among other things, the soldiers, sailors, airmen, and marines were taught how to recognize edible plants and animals, and how to live off the land. It was understood that much of this natural food would be eaten raw. A downed pilot in enemy-held jungle has little time for cooking, and cooking fires could draw some unwanted attention from the enemy. However, eating raw vegetation and raw edible animals, such as crabs, fish, and insects, could result in the acquisition of some internal additions of a more permanent nature than food. There is an intestinal fluke, endemic in Vietnam, that gains entry through the eating of raw water plants. A fluke that lives in the lungs can be obtained from eating raw crabs. Many American servicemen came home bearing these parasites in their bodies.

Flukes that live in the intestine, liver, and lungs of their hosts are more typical in appearance than the schistosomes. These flukes have flat bodies, as befits the name Platyhelminthes, and they are hermaphrodites. They have fairly large sucking disks that take in food and help them to hold on. They have fascinating, complex life cycles that involve other animals, such as crabs and fish, in addition to the usual snails. Most of these flukes are native to the Far East, but a few are parasites of domestic animals and an occasional human being in the United States.

Liver Flukes

OPISTHORCHIS SINENSIS

Throughout China, Japan, Korea, Vietnam, and Taiwan, some 19 million people, according to World Health Organization estimates, are host to a fluke called *Opisthorchis sinensis.* Generally referred to as the Chinese liver fluke (sinensis means Chinese), this worm was once so prevalent that not having it was more unusual a situation than harboring a few.

The adult worm is about 2 centimeters (about 0.8 inch) long and about 0.5 centimeters (0.2 inch) wide. They live in the bile ducts of the liver. An active Chinese liver fluke lays about 2,500 eggs a day. The eggs pass out of the host in the feces and, as is usual with flukes, must be in water to hatch. The miracidia bore into snails where they go through sporocyst, redia, and cercaria stages. The single-tailed cercariae do not bore into the primary host, but into a fish. They get in under the scales, and, eventually, get to the muscle tissue. Once inside the fish, they change into a form called *metacercaria.* A sort of capsule, made by both the parasite and the host, forms around the metacerceria. When the fish is eaten, either raw or undercooked, the metacercariae break out of their capsules and move directly to the bile ducts. In about a month, they grow to adult size and start producing eggs. Sometimes metacercariae can be ingested in contaminated drinking water.

Light infection generally causes no problems. In many areas, however, light infections are not likely to exist. As the number of worms increase, there may be irritation of the bile ducts, digestive discomfort (indigestion) that can occur at any time, not necessarily only when one has eaten, weight loss, dizziness, anemia, weakness, diarrhea, and other symptoms. As the worms continue their activities over the years, the bile ducts become thickened and the liver may become enlarged and tender.

Opisthorchis infection does not kill the host. Many people who know they have the worms, and suffer from various symptoms at one time or another, are able to go through their day-to-day lives relatively unhampered. However, moderate to heavy infections can lower the patient's resistance to other diseases.

There is no really effective treatment. Chloroquine slows down egg production. A number of other drugs were under investigation in Japan in the 1970's, and a few showed promising results in early experiments. However, as with any drug, years of testing to determine its short- and long-range safety must be carried out before it can be released for general use.

Historically, *Opisthorchis* has been more prevalent in South China than in North China. Many South Chinese regional foods, particularly those in Kwantung Province, feature strips of raw or lightly cooked fish. The problem was made worse by the practice of raising freshwater fish in ponds fertilized with night soil. This practice is not as widespread as it once was. In north China, most of the traditional recipes call for cooking fish rather than eating it raw. The incidence among north Chinese is less than 1 percent. However, in north China, the incidence among dogs and cats, many of whom are fed scraps of raw fish, is fairly high.

A curious concentration of infection exists among a small group of refugee German Jews who stayed in Shanghai during World War II and then went to the United States and other countries when the war was over. Apparently, this focus of infection stems from food fraud. Some Chinese merchants passed off infected freshwater fish as "saltwater herring" to many of the refugees. Herring (which carries no larval parasites of any kind) is customarily eaten raw, and when the fraudulent "herring" was eaten, many of the victims were infected with *Opisthorchis*, some quite heavily.

The government of China has made a great effort to reduce the incidence of *Opisthorchis*. The existence of reservoirs in wild animals, and the continued consumption of raw fish in areas where that is the custom, have hampered these control efforts.

FASCIOLA HEPATICA

Fasciola hepatica, the sheep liver fluke, is "our" fluke. This rather large, 20 to 30 millimeters (0.8 to 1.2 inches) long and 8 to 12 millimeters (0.33 to 0.5 inches) wide worm is found all over the world. It is, however, more of a problem in Europe and the Western Hemisphere than it is in the Orient.

As the name implies, it is chiefly a parasite of sheep (and also cows). However, since there is a fairly high incidence of human infection in some places, it deserves attention here, rather than in Chapter 9. It is a darkly colored worm, somewhat leaf-shaped. The worm lives in the gallbladder and bile passages in the liver. Occasionally a few are found wandering around the body cavity. Eggs pass out in the feces, and miracidia that manage to get into a snail go through sporocyst, redia, daughter rediae (sometimes), and cercariae stages. The cercariae swim around for a while, then come to rest on grass, water plants near shore, or on soil. They then change to spherical metacercariae. When the metacercariae are ingested by grazing sheep or cows they enter the duodenum (the beginning of the small intestine at the far end of the stomach). They burrow right through the duodenum into the body cavity, from where they make their way to the liver and bile passages. Occasionally they get lost along this journey, which is why they are sometimes found in the body cavity.

The worms live on bile and liver tissue. Large numbers of worms can do considerable damage to the liver over a period of time. Sometimes, one fluke can produce a painful, potentially serious blockage of a bile duct. The host may experience fever, chills, headache, abdominal pain, and, occasionally, a severe pain in the area of the breastbone. Other symptoms include diarrhea, jaundice, anemia, and digestive problems.

A drug called dichlorophenol is frequently effective. However, the drug can have adverse effects on the heart, and so must be used carefully.

In most parts of the world, *Fasciola* is mainly a problem of livestock, such as sheep and cows. Large numbers of these flukes can sicken the animal and keep it from growing as quickly and completely as it should. Livers of heavily infected animals are unfit for any food use. Human infections do occur. Areas of fairly frequent human occurrence include Cuba, Great Britain, southern France, and Algeria. Some areas of South America are also high-incidence areas. *Fasciola* usually gets into people through the eating of raw vegetables, primarily watercress. Human infection is not too common in the United States, although infection in sheep is quite prevalent. Hawaii is the only state where human *Fasciola* infections occur with any kind of regularity.

One possible reason for the low incidence of infection in the United States is that the eating of raw vegetables is not as common a custom here as it is in other parts of the world. However, the movement toward the eating of organically grown and natural foods might bring about an increase. Many organic food enthusiasts eat raw vegetables grown with natural fertilizers. Frequent consumption of such raw fare could result in having a few natural but uninvited guests in one's interior.

Intestinal Flukes

As the name *Fasciolopsis buski* suggests, this one is similar in many respects to *Fasciola*. Like *Fasciola,* it is big, but *Fasciolopsis* is bigger. The worms can be more than 7 centimeters (almost 3 inches) long and 2 centimeters (about an inch) wide. Its development in snails is similar to that of *Fasciola*, and it forms metacercariae on water plants. It differs from *Fasciola* in its fleshy color and its abode, the small intestine.

Fasciolopsis is a parasite of man and hogs. It gets into the host when raw water plants with metacercariae on them are eaten. Upon ingestion, a larval worm emerges from the metacercarial cyst and takes up immediate residence in the intestine, where it grows to adult size in about a month.

Fasciolopsis is native to central and south China, Vietnam, Laos, Cambodia, Thailand, India, Burma, Korea, and Indonesia. A number of American military personnel were infected with *Fasciolopsis* while serving in Southeast Asia. The World Health Organization estimate is 10 million so affected.

The worms are more damaging to the small intestine than many other intestinal parasites. They eat undigested food in the intestine, and probably eat some of the intestinal tissue as well. However, even if they ate little or no intestinal tissue, they would still inflict quite a bit of damage with their sucker disks. Inflammations, ulcers, and abcesses in the intestine occur frequently in *Fasciolopsis* hosts.

The worms can cause digestive distress. Ironically, the distress is sometimes the cause of getting more worms. People in endemic regions customarily eat raw water chestnuts to relieve the digestive distress, a bit of home treatment that can bring in more larval worms.

Treatment with a drug called tetrachlorethylene has been successful, although control and prevention is difficult because of the high incidence of pig infections. In many parts of the Orient, pigs are allowed to roam free and forage food for themselves. This is an economical practice, but it also means that pig excrement, containing worm eggs, can get into just about anything, including bodies of water where they are snails. Cooking water plants would kill the metacercariae, but the eating of these plants raw has been a custom too long for it to be easily or quickly discouraged.

Lung Flukes

There are several species of flukes that can make homes for themselves in the lungs of humans and other animals. *Paragonimus westermani* is one of the more widespread of the lung flukes. While it is primarily a parasite of the Far East, it is also found in South America and Africa. It also occurs on many of the islands in the South Pacific.

Paragonimus passes through two intermediate hosts in its development. After the cercariae emerge from the snail, they burrow into a freshwater crab or crayfish. If infected crabs or crayfish are eaten raw, the live metacercariae are ingested.

The larval worms get to their final home in a rather direct way. They burrow through the wall of the duodenum and break out into the abdominal cavity. They then penetrate the diaphragm, the tough sheet of muscle that separates the chest from the abdomen, and go into the chest cavity, and from there to the lungs in a month or so.

The worms exist in the bronchial tubes, enclosed in a kind of cyst or capsule. Eggs are released into the bronchi and are coughed up. They can exist the host in two ways: they can be spit out, or they can be swallowed, in which case they will eventually exit in the feces.

Coughing is a characteristic symptom of *Paragonimus* infection. The coughing brings up phlegm, which is usually bloodstained and brownish in color from the eggs. There may be pain in the chest area accompanied by fever.

The consumption of such oriental food delicacies like "drunken crab" has served to spread this parasite. Drunken crab is prepared by soaking the crabs in vinegar, wine, or saltwater. Such treatment might kill the crabs, or at least put them into a stupor, but it seems to do nothing to the metacercariae, at least for several hours. Drunken crab was, and still is, sold in street stalls in many cities in the Orient. In some parts of Korea, crab juice, prepared by crushing crabs, is considered to be a cure for measles. Children are often infected by this home remedy. During World War II, some American troops were infected. Many of these infections were the result of impromptu picnics out in the field, in which crabs or crayfish were the featured food.

A drug called bithionol is an effective treatment. This substance was once used as an ingredient of shampoos and cosmetics. Such use is no longer allowed. It was found the bithionol caused skin rashes. Not surprisingly, skin rashes are one of the side effects of treatment with the drug.

Paragonimus infection can be prevented by cooking crabs. However, the fondness for drunken crab and other raw crab and crayfish foods will probably help to keep *Paragonimus* around for a while.

7.

Protozoa

The thoughts of yards and yards of tapeworms or tangles of foot-long ascarids in one's intestine, or of slippery flukes slithering through one's liver is revolting to many, if not most, people. Much of the revulsion to worms, such as ascarids, comes from the fact that they are big enough to be clearly seen. However, the degree of damage a parasite can do to its host is not directly proportional to its size. As pointed out earlier, millions of people have been host to large worms, such as *Ascaris* and tapeworms, and have never suffered any ill effect. Far more serious harm is likely to come from organisms such as protozoa, bacteria, and viruses that are much too small to be seen without a microscope. Because these organisms are tiny, and their presence cannot be determined by unaided vision, few people react with disgust and revulsion at the mere thought of them. Tiny organisms, however, cause horrible disease such as bubonic plague and rabies, a fatal disease that kills its victims slowly, agonizingly, and painfully.

The protozoa are considerably bigger than bacteria and viruses, but they are exceedingly tiny compared to an ascarid or a large tapeworm, such as *Dibothriocephalus.* Most protozoa are free-living. Many parasitic protozoa, such as *Entamoeba gingivalis* (page 51), are more like commensals than like parasites; most hosts never know they are there. There are, however, a few protozoa that cause some of the most important and horrible diseases ever to afflict mankind.

Most protozoans are made up of one cell. They are probably very much like some of the earliest forms of life to evolve on the Earth billions of years ago. They are found just about everywhere as free-living organisms and as parasites. There is hardly a living thing on this planet that does not harbor some kind of protozoan on either a casual or permanent basis. For example, sporocysts and rediae of many species of schistosomes are commonly parasitized by protozoans. Many roundworms harbor them, and there are few insects that are not host to one or more species of protozoa. Some are necessary to the survival of the insects; recall the tiny protozoa that live in the intestines of termites and digest the wood for their insect hosts.

Protozoa are considered to be animals by most biologists, although some are very plantlike in their behavior. Some biologists think protozoa are neither plant nor animal, but should be placed in their own kingdom called Protista. This controversy is of no concern to us in the context of this book. The organisms that cause African sleeping sickness and malaria, for example, do the damage no matter what they are called.

Plasmodium Spp.

While you read this, somebody somewhere is suffering from the racking chills and burning fever of malaria. Malaria has long been one of the great scourges of the earth. It probably originated in Africa and spread from there to other parts of the world, either through voluntary migrations of people or the involuntary migration that was African slavery. We probably inherited the disease from our prehuman ancestors.

Although malaria has most likely existed since before man evolved, it has not always been a problem. There is nothing in ancient Egyptian writings to indicate it was troublesome in the earliest years of Egyptian civilization. By the time of Alexander the Great, however, malaria was widespread in Greece. Indeed, the campaigns of Alexander probably helped to bring malaria to Greece and other parts of the Mediterranean. Alexander himself may have died of malaria, but there is no proof of that.

From Greece it spread to other parts of Europe, including present-day Italy. The Etruscans, who were in Italy before the Romans, did not suffer from malaria. Nor was it a big problem in Italy at the height of Roman power and influence. However, as Rome declined and everything started to fall apart, including the efficient drainage systems developed by the Romans, swamps and marshes appeared, and the mosquito population increased.

Of course, the Romans did not know that the malaria organism is carried by mosquitoes; that has been known less than one hundred years. Mosquitoes need water if they are to breed, and swamps and marshes provide them with more than enough. It was (and still is in some places) believed that malaria was caused by damp night air—hence the word malaria, which comes from words that mean "bad air" in Latin and Italian. On the other hand, many Africans had a pretty good idea of the relationship between mosquitoes and malaria. In one of the tribal languages used in Tanzania, the word for mosquito, *mbu*, is the same as the word for malaria.

Even in the Western world, malaria was not always referred to by that name. There are different kinds of malaria, each one with its own particular set of characteristics. People in various areas gave their own names to this disease long before it was known that mosquitoes had anything to do with it. In some parts of the world these names are still used.

In the American South, malaria was generally called "chills and fever." In Britain and the United States, it used to be called "ague." Looking through newspapers and magazines of the latter part of the nineteenth century will reveal many advertisements for patent medicines that were claimed to be cures for "ague" and other afflictions. Other names include tropical fever, marsh fever, coastal fever, blackwater fever, and paludism.

The Greek physician Hippocrates wrote about "miasmic fevers." These were thought to be caused by "noxious emanations from soil and earth." He classified miasmic fevers into three types: quotidian, tertian, and quartan. In all probability, some of the fevers he so named were malaria. The terms refer to fevers that afflicted the patient at regular time intervals, a description that fits malaria. In any case, the terms persist into modern times, and they are still used to describe certain kinds of malaria. However, the trend of late has been to refer to the disease by the name of the parasite that causes it.

Human malarial parasites were first seen in 1880. From that time on, more and more about malaria was discovered, and by 1900 the steps of development in mosquitoes had been worked out. However, one kind of malaria *(Plasmodium ovale)* was not discovered until the 1920's, and some stages of the disease in human hosts were not understood until the 1950's. There is still a great deal that is not known about some stages of the parasite's life in the human host.

Malaria is caused by a genus of protozoan called *Plasmodium*, which belongs to a class of protozoa called Sporozoa. All the protozoa in this class are parasites. Unlike most other protozoa, which can move about quite well, Sporozoa are not speedy movers. They are usually transported in the body of the host by the bloodstream and by other natural movements of the host's body.

The parasites can be seen by examining the blood. A little blood is taken from the patient and smeared on a slide in either a thick or thin film, depending on what stages are being sought. The blood smear is stained with dyes so the parasites can be clearly seen with a microscope. They are most easily seen when they are inside red blood cells. There are techniques for detecting them in other parts of the body.

Plasmodia appear in an amazing variety of shapes and forms. One of the most difficult parts of the education of medical technologists is learning to recognize the various stages of the different types of plasmodia. Sometimes they appear in "signet ring" shapes. At other times, they are irregularly shaped, somewhat like an amoeba. Knowing which one is which is extremely important. Treatment, both short- and long-range, is different for each type of malaria.

LIFE CYCLE

The Sporozoa have very complicated life cycles involving

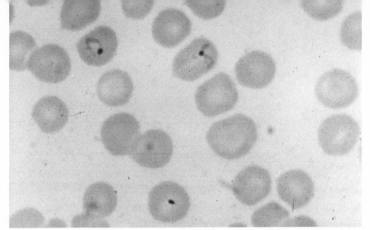

Plasmodium vivax. *The round objects are red blood cells. Two "signet ring" stages can be seen in the upper part of the picture. A trophozoite is at lower center. Magnification is about 1125X.*

THE CENTER FOR DISEASE CONTROL

various stages that alternate in reproducing sexually and asexually. There are four species of *Plasmodium* that cause malaria in people: *P. vivax*, *P. falciparum*, *P. ovale*, and *P. malariae*. The life cycles of all four species are similar, but there are certain differences. The differences in the life cycles have a bearing on the severity of the disease.

The basic life cycle: Following the life cycle of a *Plasmodium* is a bit easier if we start with the intermediate host's bite. The intermediate hosts of *Plasmodium* are various species of *Anopheles* mosquitoes. These are not the common variety of mosquito that you would be likely to slap at once or twice during a hot night in Iowa. The genus of mosquito most likely to bite in the United States is *Culex*, although species of *Anopheles* do exist in the United States, particularly in the southern states. Malaria was once a health problem in the United States, and not just in the South. The occasional *Anopheles* you might see is a legacy of what once was here.

The best way to tell an *Anopheles* from a *Culex* with a quick glance is how they stand. A *Culex* sits horizontally. An *Anopheles* seems to stand on its head.

When an *Anopheles* mosquito bites a person with malaria, it may take in some plasmodia that are in the separate sex stage. These stages are called male and female *gametocytes*. The gametocytes are taken up into the mosquito's stomach and go through some changes. The female and male gametocytes mature into another form called *microgametes*. A male and female microgamete fuse and become a *zygote*. This zygote changes into one of the few forms of *Plasmodium* that moves around a little.

The zygote, called an *oökinte* in its active stage, bores into the stomach wall, rolls itself up into a little ball, and develops anew. Through a series of cell divisions, the oökinte changes into many slender, threadlike *sporozoites*. The sporozoite is the form that can infect a new host when the mosquito bites.

The sporozoites break out of the stomach and are scattered throughout the body of the mosquito. Those that happen to enter the mosquito's salivary glands are the ones that might make trouble for the next person the mosquito bites. When the mosquito bites, she shoots in some liquid from her salivary glands to keep the blood from clotting in her mouth parts. The sporozoites travel in the salivary gland liquid.

When the sporozoites get into the bloodstream, they move into the liver. This invasion of the liver takes place very quickly, usually within forty minutes after the mosquito bites. In the liver, the parasites go through a complicated development. The result of the development is the form of *Plasmodium* that wreaks the most havoc in the host.

The form of the parasite that develops in the liver leaves

that organ and goes into the bloodstream. The parasites enter red blood cells and go through another complicated series of developmental stages. These are different in each species, but the basic process is the same for all: the parasite enters the red blood cell and goes through several reproductive stages, resulting in a form of *Plasmodium* called *merozoites.*

The merozoites burst out of the red blood cells and go into more red blood cells and make more merozoites that go into more red blood cells and so on. This cycle can go on for some time. Eventually, some merozoites become new gametocytes and, when these are taken up into an *Anopheles* mosquito, a new cycle is started.

SYMPTOMS

The plasmodia can stay in the liver for periods of time ranging from two weeks to years before breaking out into the bloodstream. How long this period in the liver, called the

A schizont phase of Plasmodium vivax. *The schizont goes through changes to become many merozoites, each of which can infect another red blood cell.* THE CENTER FOR DISEASE CONTROL

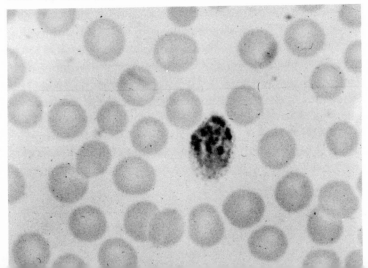

"incubation period," lasts depends on many things, such as the species of *Plasmodium*, previous infections, and other factors that are still not fully understood. The first symptoms can begin as early as three weeks after the mosquito introduced the sporozoites. The patient may have a headache, loss of appetite, aversion to strong light, muscle aches and pains, nausea, and listlessness. The symptoms occur toward the end of the incubation period. These vague aches and pains are nothing compared to what usually follows. Sometimes there are no primary symptoms at all.

When the red blood cells break apart, releasing newly formed merozoites into the blood stream, the patient goes into what is called a "paroxysm." The paroxysm starts with a sudden chill that causes the patient to shake uncontrollably. The patient may go into a "rigor"; a stiffening of the muscles and clenching of teeth. Although the actual body temperature will be a bit above normal at the start of the paroxysm, the feeling of extreme cold is very real to the patient. Blanket piled upon blanket does nothing to relieve the hacking, shaking, shuddering chill.

Following the chill, the patient goes into the hot phase. The body temperature climbs steadily, and if the patient has been under any blankets, they will be thrown off. There may be other symptoms, depending on the kind of malaria the patient has. The patient's skin is usually flushed during the fever, and he may feel pain in other parts of the body. Sometimes the patient becomes delirious during the fever, babbling incoherently and perhaps flailing about, tossing and turning. Nausea, with or without vomiting, can also occur.

The hot phase is generally followed by profuse sweating. With the onset of sweating the patient usually begins to feel better. Usually he is completely exhausted by the time the sweating stage stops and will feel like sleeping. When the patient wakes up, the paroxysm can be considered to be over, and he feels much better until the next one starts.

The paroxysm is brought on by the release of all kinds of material from the burst red blood cells. This material includes not only the merozoites, but also the remains of the red blood cells and various waste materials produced by the parasites. The release of all this material into the blood is like getting an injection of poison. Of course, the destruction of large numbers of red blood cells does the patient no good. Anemia is, indeed, one of the long-range effects of malaria. Other long-range effects include damage to the liver, spleen, and other organs.

A series of paroxysms is called an attack. The length of the attacks, how many years the patient can expect to have to endure them, and how often they recur are matters that vary with each species of *Plasmodium.* The length of the chills, fever, and other symptoms of the paroxysm also varies with the species. With some species of *Plasmodium*, the disease is over and done in less than a year; another can linger for thirty years or more.

One of the most intriguing things about malaria is that paroxysms occur at more or less regular intervals. The time between paroxysms is one way to diagnose which malaria the patient has. The early paroxysms tend to be irregular, but as the parasites become established, their activity becomes more and more regular. It is believed that the early fever episodes synchronize the plasmodia in some way. The regularity makes the attacks more severe, as large numbers of parasites and other materials are released into the blood at one time.

P. vivax

P. vivax is considered to be in the mid-range of severity and general range of effects. The preliminary symptoms that occur at the end of the incubation period are usually mild compared to those of some other kinds of malaria. The incubation period of *P. vivax* is generally about two weeks, although in some situations it can be as long as a year.

P. vivax chills last about fifteen minutes, but tend to become longer as the disease progresses. The hot stage can go on for two to six hours. The paroxysms occur in a forty-eight-hour cycle, a circumstance that caused this form of malaria to be called "tertian," that is, "third day." Attacks can go on for weeks or months, gradually becoming less severe as the paroxysms become more irregular, until they fade away.

Relapses, the return of attacks, do occur with *P. vivax*. Relapses are more likely to occur with *P. vivax* than with *P. falciparum*, but less likely than with *P. malariae*. In the period between attacks, the parasites stay in the liver.

About 2 percent of the blood cells are affected by *P. vivax* infection, a figure that is considerably less than in *P. falciparum* but more than *P. malariae* and *P. ovale*.

P. ovale

The symptoms and overall effects of *P. ovale* are similar to *P. vivax*. The paroxysm cycle, for example, is also forty-eight hours. *P. ovale* produces a slightly milder disease than *P. vivax*. Relapses are less likely to occur and don't last as long. *P. ovale* infection is so mild it has been used to treat another disease, neurosyphilis, one of the later stages of the very dangerous disease, syphilis. In this later stage, the nervous system is affected. High fever sometimes has a beneficial effect, and there was a time when syphilis patients were deliberately infected with *P. ovale* and other forms of malaria. Often, however, the treatment did not work because the fever produced by *P. ovale* infection was not high enough. Antibiotics rather than malaria are the usual treatment for syphilis today.

P. malariae

This one seems to have a preference for older red blood cells. This peculiarity may explain the generally slow course of this type of malaria. The incubation period is usually about four weeks, but it can be years. Paroxysms occur in a seventy-two-hour cycle, ergo the term "quartan," and are generally a little more severe than those of *P. vivax.* The chill stage lasts longer and may be accompanied by vomiting. The hot stage is also longer than that of *P. vivax* and the patient may also be plagued with vomiting, diarrhea, and severe headaches. Sometimes the patient goes into a complete collapse during the sweating stage.

P. malariae is the most likely of the human malarias to relapse, and when it does, the duration of the relapse is long. The duration of the primary infection is the longest of all the malarias—thirty years is not uncommon.

P. falciparum

This parasite is considered to be the most dangerous by many authorities. Although its duration is short (six months to a year and half) and the chances of relapse are the least of all the malarias, the chances of fatal complications during the disease are the highest in this form.

SUMMARY OF THE FOUR TYPES OF HUMAN MALARIA

	other names	cycle	incubation period	average concentration of parasites in blood	severity of attack
P. vivax	benign tertian	48 hours	usually 14 days; can be 6–12 months	moderate	mild to severe
P. malariae	quartan	72 hours	usually 28 days; can be years	low	severe
P. falciparum	malignant tertian subtertian estivo-autumnal	48 hours	9–14 days	very high	mild
P. ovale	benign tertian ovale	48 (49) hours	17 days; perhaps years	low	mild

SUMMARY OF THE FOUR TYPES OF HUMAN MALARIA (Cont'd)

	length of paroxysm (hours)	length of attack (weeks)	tendency to and length of relapse	length of disease (years)	chance of fatality; overall damage to body
P. vivax	8–12	2–4	strong tendency; long duration	2–5	liver damage; low to moderate
P. malariae	8–10	4–8	very strong tendency; very long duration	1–30	low
P. falciparum	16–30 or more	2	weak tendency; short duration	0.5–1.5	high; brain involvement
P. ovale	8–12	3	strong tendency; long duration	2–5	low

The incubation period is ten days to two weeks. The pre-paroxysm symptoms are similar to those seen in the other forms. The chills are not as severe as in the other forms, and sometimes there is hardly any chill at all. The fever occurs in a variety of patterns. Sometimes the fever continues for a long time, at other times it comes and goes at frequent intervals. Although *P. falciparum* does have a cycle—forty-eight hours—there is often a wide variation in the time between paroxysms. Attacks generally run their course quickly, seldom lingering more than three weeks.

What *P. falciparum* lacks in severity of symptoms, it more than makes up for in some of its long-range effects. One of the biggest dangers of *P. falciparum* is the blocking of blood vessels by collections of diseased red blood cells. The blocking of blood vessels and allergic reactions to the parasites can bring on complications in the brain, kidneys, gastrointestinal system, heart, and other organs. Any of these complications could result in the death of the patient.

TREATMENT

There is an arsenal of drugs available for the treatment of malaria. The choice of drug depends on the type of *Plasmodium*, how far along the disease is, the general condition of the patient, what stages the physician wants to kill, and much more.

The first drug to be used with any degree of effectiveness was quinine. It is still used. Quinine is obtained from the bark of the cinchona tree. According to one story, the tree is named after the Countess d'El Chinchon, wife of the Spanish viceroy of Peru, who, in 1638, was cured of malaria by eating the bark of the tree.

Quinine and similar compounds were the only an-timalarial drugs available until the 1920's. Since that time many more drugs have been manufactured. No one drug is effective against all the plasmodia, nor is one drug effective against the parasite in all its many stages.

One of the big problems with treating malaria is that the plasmodia develop resistance to one or more drugs after a while. A drug will be effective for a time, killing plasmodia by the millions. However, the drug doesn't kill all of them. Some are able to resist the drug and survive, then produce more plasmodia that will inherit the resistance.

There are three basic ways to treat malaria: suppressive therapy, clinical cure, and radical cure.

Suppressive therapy is the giving of progressive small amounts of drugs, so that the parasites are killed as soon as they enter the bloodstream. A drug called quinacrine (trade name, Atabrine) was used for this purpose during World War II. Taking Atabrine pills became a familiar routine to American military personnel serving in the South Pacific. The drug is actually a dye, a fact soon discovered by natives of many South Pacific islands. In *Tales of the South Pacific* James Michener wrote of how, when given the drug, the natives would hide the pills in their mouths rather than swallow them. Later, they would retrieve the pills and use them to dye straw for making grass skirts, which they sold to the Americans.

During the Vietnam War, a number of drugs were used for suppressive therapy when it was found that one drug was not totally effective. Among the suppressive drugs used in Vietnam were chloroquine, Dapsone, and pyrimethamine.

Generally, larger quantities of the same drugs that are

used for suppressive therapy are also used for clinical cures. A clinical cure involves killing the parasites in the red-blood-cell stages of the disease. A clinical cure does not affect any parasites that may be in the liver. Although a clinical cure can be effective, there is always the possibility that the disease might come back when the parasites in the liver become active again.

A radical cure involves killing the parasites in the liver and elsewhere as well as those in the bloodstream. Several drugs are available for radical treatment and they are often effective, although no one seems to know just how they work. Others are known to retard the reproduction of the parasites. Drugs used for radical treatment must be used very carefully.

Malaria drugs have a way of making comebacks after they have long been ignored. For example, a drug called colchicine was widely used during World War II. As newer drugs were developed, colchicine fell into disuse. However, when *P. falciparum* developed resistance to a number of drugs, colchicine was again widely used, particularly in combination with quinine.

In addition to the giving of drugs to kill the parasites, the patients need general therapy to make them feel better and to recover from the debilitating effects of the disease. Anti-fever drugs, such as aspirin, are given during the hot phase. Sponging with cold water also helps to relieve the fever. Anemia is treated with vitamin and iron therapy. Fluids are sometimes given intravenously to help restore fluid balance.

FACTS AND FIGURES

Today, malaria is mostly limited to tropical areas. According to World Health Organization estimates, more than one billion people live in areas where they are at risk of getting malaria. Some 177 million have malaria, and about one million a year die from the disease. These figures sound high and, indeed, they are, but they are considerably less than they were not too long ago. In the 1940's, more than 300 million people had malaria, and 3 million people a year died from it. But there were fewer people living in areas of risk in the 1940's, since the earth's population was half what it is today, so those figures represent a much larger percentage of people. The dramatic and frightening increase in world population over the past thirty to forty years is due in no small measure to the elimination of malaria as a major cause of death in many parts of the world.

After World War II, there was a concerted effort to eradicate malaria, and although the disease has not been eradicated, it has been substantially reduced in many areas. The success of this effort was made possible mainly by insecticides such as DDT. The results of the antimalaria campaigns have been dramatic in many areas. New problems, however, have been created. DDT is a lingering substance, a quality that enhances its effectiveness in killing mosquitoes. That same quality, however, poses a great danger to all forms of life on earth, even at a great distance.

Formerly, in some areas the entire economy was adjusted to the inevitability that a certain number of people would die of malaria, and that large numbers of infants would die before they reached the age of one. DDT and other antimalaria measures and the reduction of infant mortality rates caused sharp increases in population in many areas. Food shortages, unemployment, and overcrowding were frequent results, chronic examples of how efforts intended to improve the quality of life can bring on new problems.

Malaria had been entirely eliminated from the United

States, but in recent years, new cases have appeared among immigrants and travelers to areas where malaria is newly resurgent. It has been almost entirely eliminated from southern Europe, and has been largely eliminated in Puerto Rico, Jamaica, many of the Carribbean islands, and from most areas in Brazil, Argentina, Venezuela, and Ecuador.

Malaria today is found mostly in tropical and subtropical areas. *P. falciparum* is almost entirely limited to tropical regions. In some of the cooler areas where *P. falciparum* is found, most attacks occur in the summer and autumn, hence the name *estivo-autumnal*, which is often applied to this form of malaria.

It is still a good idea to take small does of antimalarial drugs as suppressive therapy if you are planning to travel to tropical regions where malaria still exists, and especially if you plan to stay there for a while. The use of such drugs is one of the ways malaria can be kept under control. Killing mosquitoes is the other obvious way to control malaria. Of course, it is impossible to kill every single anopheline mosquito, but control is possible. The use of insecticides is still the most effective way to control. In some areas, fish that eat mosquito larvae have been introduced into ponds, marshes, and so on. General cleanliness and orderliness are important in mosquito control. An old tin can can hold enough water to provide breeding for hundreds of mosquitoes. Protecting malaria victims from exposure to mosquitoes is very important. That, however, is not always easy, since the required materials—screening, netting, and insect repellents—are sometimes too expensive.

In Africa a sort of natural malaria control evolved millions of years ago. This control is sickle-cell anemia, which is a serious disease, resulting in painful attacks that come and go over the years. The disease is not caused by any organism, but is inherited, caused by a gene that alters the hemoglobin, the substance in the red blood cells that combines with oxygen and carries it to cells in the body. One of the characteristics of the disease is that the red blood cells assume a crescent, or sickle, shape, hence the name sickle-cell anemia. Trouble comes when masses of these oddly shaped cells jam up and block blood vessels.

The curious thing about sickle-cell anemia is that malaria organisms do not develop in affected red blood cells. The disease was passed on and those afflicted with it tended not to die of malaria and live long enough to have children. Many black Americans have sickle-cell anemia.

The 1970's resurgence of malaria was a vivid reminder that the disease will not go away by itself. Constant effort and lots of money in the right places are needed to keep this disease under control. If antimalarial efforts are neglected, as indeed happens when there is strife, war, and other nonproductive activity, the mosquitoes and parasites will go right ahead and do what millions of years of evolution have always directed them to do.

Amoebas

ENTAMOEBA HISTOLYTICA

Mention has been made of *Entamoeba gingivalis* (page 51), the commensal parasite that lives in human mouths. There are at least three other species of amoeba that live as probable commensals in various parts of the human anatomy. One that is definitely a parasite is *Entamoeba histolytica*, but this amoeba seems to exist in several varieties or strains. These might in actuality be several species, but this has not been proved. Different strains of *E. histolytica* are affected by different drugs. There is no completely dependable way of telling one strain from the other.

LIFE CYCLE

Active *E. histolytica* are typically amoeba-shaped, meaning that they have no shape. Ranging in size from 10 to 60 micrometers in diameter, they move by extending their pseudopods, as do most other amoebas. They live mostly in the human large intestine, where they work themselves into the mucosa and feed on red blood cells and bits of tissue. They reproduce by cell division. From time to time, the amoebas roll themselves into a sphere and secrete a hard protective substance that covers the cell. This is a cyst, and it is the infective stage of the amoeba. Cysts pass out of the host in the feces. When the cyst is ingested by a new host, the protective coating is digested away. The enclosed amoeba immediately divides into eight new little amoebas, which move down to the large intestine and take up residence. The actively moving stages of the amoebas are called *trophozoites.*

SYMPTOMS

The amoebas give off enzymes that dissolve the tissue of the intestinal mucosa. The amoebas then ingest the red blood cells, tissue bits, and other materials released by this process by engulfing them with their pseudopods. All this activity does the intestinal mucosa no good. Ulcers form in the intestinal mucosa, and the damage can be considerable. Some-

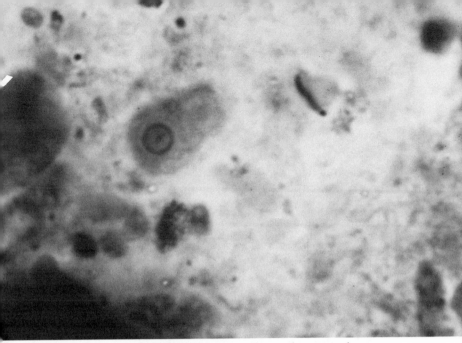

A trophozoite of Entamoeba histolytica. *The trophozoite is the moving, active stage of the organism.*

**YALE UNIVERSITY SCHOOL OF MEDICINE
DEPARTMENT OF COMPARATIVE MEDICINE.
PHOTO BY ROBERT SULLIVAN**

times the amoebas get into the bloodstream and are carried to other organs that provide suitable living conditions for them. The liver is the most frequent nonintestinal living site. Large numbers of amoebas in the liver can cause abcesses and other damage.

Occasionally there can be so many of them in the large intestine that the intestine is perforated. Sometimes they cause appendicitis. Other times, they might erode away a large blood vessel, causing a massive hemorrhage. Over the years, the scar tissue forms in the large intestine, resulting in the accumulation of thick, fibrous growths in the intestine.

Symptoms vary widely, depending on the condition of the

patient and, in all probability, the strain of amoeba. The most obvious symptom is dysentery, which is characterized by painful, severe diarrhea. General amoebic infection is often called amoebic dysentery, but that is a misnomer since dysentery does not always occur in amoeba hosts. The dysentery can be so bad that it is called acute dysentery, in which the victim has almost constant bowel movements that usually contain blood, mucus, and bits of intestinal tissue. There is severe pain in the abdomen, dehydration, toxic reactions, high fever, and prostration.

Although the victim of dysentery suffers miserably, he is not a direct public health menace at this time. There are amoebas in the dysenteric stool, but they are noninfective trophozoites. Only the cysts are infective.

Infections that produce no symptoms are quite common. Asymptomatic infections are most prevalent in temperate areas such as the United States. Although the person may suffer no symptoms, he or she is a carrier and can spread the disease. Some people who accidentally ingest the cysts can become very sick, and others can be carriers with no symptoms for years and then become sick with dysentery. Carriers might have some symptoms, such as vague intestinal discomfort, but symptoms such as those can be due to any number of things, such as mild digestive upsets from eating too much, virus infections, or not enough sleep.

The disease can become chronic. Some victims of chronic amoebiasis have dysentery attacks that come and go. There might be constipation and general gastrointestinal discomfort between bouts of dysentery. The liver is usually enlarged, and there is damage to the large intestine, ranging from mild to severe.

Amoebiasis in the liver can be a very serious matter and

is frequently fatal. The liver becomes enlarged and tender. Abcesses form, and the functioning of the organ is impaired. Amoebas can also become established in the lungs and brain.

TREATMENT

Methods of treatment vary with the severity of the disease. Of course, the main idea is to kill the amoebas, but care must be taken not to make the treatment so drastic that a weak patient is put into danger. A group of drugs called iodides are effective but cannot be used if there is extensive liver damage. There are other drugs, including antibiotics, which are effective. The cure rate is fairly high, but there is also a high rate of relapse. It is not easy to kill all the amoebas with the medication, and there are resistant strains for which new drugs or combinations of drugs are needed.

A very effective drug for acute amoebic dysentery is emetine hydrochloride. It is given by injection. However, if given for too long a period of time, it can have toxic effects on the patient.

General supportive therapy, such as rest, good food, and restoration of fluids, is also important. The effects of amoeba infection are often aggravated by malnutrition.

FACTS AND FIGURES

The World Health Organization estimates that 10 percent of the world's population has amoebiasis, either symptomatic or nonsymptomatic. It is found mostly in tropical areas and, as might be expected, is a disease associated with filth and squalor and lack of sewage-disposal systems. It has, however, occurred in some very affluent places. Epidemics of it have been traced to expensive, exclusive restaurants in big

Western cities. Amoebic dysentery is not to be confused with what is often called "Montezuma's revenge." That affliction, which affects tourists in Mexico and other tropical and subtropical areas, is caused by bacteria or viruses rather than amoebas. Montezuma's revenge is uncomfortable and inconvenient, but it passes quickly and usually does the victim no permanent harm. One attack often confers immunity. Amoebic dysentery can also be contracted in Mexico and other places popular with tourists, but amoebiasis seldom ruins one's vacation. The symptoms don't start until weeks after exposure (one to fourteen weeks), or may not develop for years—or never, if the host becomes a carrier.

Prevention is best accomplished by general cleanliness. The same measures used to control intestinal worms can be applied to controlling *Entamoeba.* Food handlers should be instructed, indeed indoctrinated, to wash their hands thoroughly and frequently, especially after going to the toilet. Tourists should avoid eating fresh, uncooked fruit and vegetables in areas where amoebiasis is endemic. Boiling water before drinking is also helpful. Freezing for short periods does not kill cysts, so ice should be made only from boiled water in endemic areas. In the field, iodine tablets kill cysts, but one has to put up with strange-tasting water.

NAEGLERIA

Naegleria is a genus of amoebas that act like amoebas in one generation and like flagellates, another class of protozoa, in another generation. This amoeba has caused increasing concern among the United States public-health officials in recent years.

Not too much is known about its life cycle, but apparently

it can be free-living or parasitic. The parasite enters the noses of people swimming in freshwater and brackish water. Swimming pools have also been sources of infection. Once the amoebas get into the nose of the host, the events that follow are swift and tragic.

The amoebas move from the nose up into the frontal parts of the brain and from there to the spinal cord. Symptoms start a day after exposure. There is headache and fever, followed by nausea and vomiting. The patient's sense of smell and taste may become altered. The patient begins to behave in an irrational way, then goes into a coma and dies. The patient is dead from three to six days following the entry of the amoebas into the nose.

The disease caused by the amoeba is an inflammation of the brain and spinal cord and, as such, is a *meningitis.* It is a frightening disease, and there is no treatment for it.

The disease was first described in Australia and North America in the 1960's. Since that time, it has been reported in the United States and Europe. In the late 1970's there was a disturbing increase in the number of cases in the United States, particularly in several Southern states—North Carolina, Georgia, Texas, and Florida. The amoeba has also been reported in California and New York, as well as in other states. There may be other species of amoeba that cause the same or a similar disease. However, these other amoebas may be varieties of *Naegleria.*

The proliferation of home swimming pools may be a factor in the increase of *Naegleria* disease. Many of the cases have been traced to backyard pools. Concentrations of chlorine at ten parts per million, a level called "super-chlorinated" that kills algae, will not kill *Naegleria.*

Although there seem to be more of them in small bodies of water, such as ponds and "swimming holes," than in large lakes, limiting swimming to large freshwater lakes is no guarantee of safety. They seem to exist in greater profusion in warm, still water than in cold, moving water. Although not thought to be a problem in saltwater, they do exist in brackish (semi-salty) water in inlets close to shore. The possibility that they may exist in seawater cannot be completely discounted yet.

Authorities advise swimmers to avoid diving or jumping into suspected waters. Diving serves to force water up into the nose and may be just what the amoeba needs to get into the brain and, ultimately, the spinal cord. Wearing nose clips might be of some help. Frequent blowing outward to clear the nose of water might also a preventive measure.

Most cases occur among children in the summer, simply because children spend more time in the water and dive and frolic around more than adults do. If you are in an areas where there are reports of *Naegleria* disease, the best thing to do is stay out of the water.

Naegleria is an example of an occasional parasite that is not well adapted to the host, nor is the host well adapted to it. It obviously does not "belong" in the human brain and spinal cord, as the dramatic, tragic response of the host to its presence demonstrates.

Flagellates

Flagellates belong to the protozoan class Mastigophora and are characterized by having a *flagellum*, a flexible whiplike extension that serves much like an oar to move the organism through the water.

Flagellates live in all kinds of water. New species are discovered all the time. They have been found under the ice in polar waters, in the warmest tropical sea, and in every body of fresh water of whatever size.

Thousands of species of parasitic flagellates live in just about every species of animal on this planet. The human animal, unfortunately, is no exception. Several flagellates are so pervasive that they prevent the settlement of large regions in Africa and South America. Some cause diseases more disfiguring and horrible than elephantiasis. A book that covered all the flagellate diseases of mankind in detail would be a thick one indeed. We will consider a few of the more important and widespread of these afflictions.

Trypanosomes

African sleeping sickness is caused by a kind of flagellate called a trypanosome. There are several species of trypanosomes that cause different diseases in different parts of the world, and the one that causes African sleeping sickness is *Trypanosoma brucei.* There are varieties or subspecies of *T. brucei* that cause slightly different forms of the disease in different parts of Africa.

Trypanosomes are actively moving single-celled animals. The cell is from 15 to 20 micrometers long and somewhat flattened on the sides, giving it a sinuous appearance. The flagellum proceeds from the front of the cell and then follows a sort of side flap called the *undulating membrane.* The overall appearance is that of a free-flowing, diaphanous creature. Trypanosomes are fascinating to look at, but they are nothing you would care to have in your body. Unfortunately, that is just where far too many of them happen to be in too many people.

Gambian Trypanosomiasis: Mid-African Sleeping Sickness

This form of sleeping sickness is caused by the variety of trypanosome called *Trypanosoma brucei gambiense*. Some biologists insist that it should be called *Trypanosoma gambiense* and that the trypanosome that causes the other type, *Rhodesian sleeping sickness*, should be called *Trypanosoma rhodesiense*.

LIFE CYCLE

Humans are the main hosts of this parasite. It is not known to establish in wild animals, therefore, reservoirs among wild animals are not a problem with this variety of trypanosome. The intermediate host is the tsetse fly of the genus *Glossina*.

The tsetse is a bloodsucking insect. Like most other insects with similar tastes, it shoots saliva into its sucking mouth parts to keep the blood from clogging them up. The trypanosomes ride in on the saliva stream into the bloodstream. The cycle is continued if a fly bites a human host and sucks up some trypanosomes with its meal. The trypanosomes go through a series of changes in the tsetses, but the changes are simple compared to the saga of Sporozoa in mosquitoes and other insects. Sometimes the trypanosomes have no chance to go through any changes in the fly. A fly might get brushed off in mid-bite by an understandably annoyed host. The fly, still hungry, will immediately go to another nearby person to complete the meal. In so doing, the fly just might send in some trypanosomes picked up from the previous "bitee" that are still in the fly's mouth parts.

SYMPTOMS

The tsetse is capable of inflicting a vicious bite, and the bite site often festers and forms a scabby sore. Other than that minor annoyance, the host will feel fine. The sore, called a "trypanosomal ulcer" will disappear. However, the immediate area around the sore is filled with trypanosomes.

Weeks and months may go by before the patient begins to suffer any symptoms. During this time trypanosomes can be found in the blood. In many instances, the trypanosome infection will not take, and the host will never get sick, but he can be a carrier. However, if the trypanosomes reach the lymphatic system, the patient will start to feel uncomfortable.

When the parasites get into the lymph nodes, the patient has fever, headaches, weakness, and general listlessness. During the lymph invasion phase, the lymph nodes, usually those in the neck, might swell. There is sometimes nausea and a general stiffening of the muscles and heavy sweating

Trypanosomes, which cause African sleeping sickness, in human blood. **THE CENTER FOR DISEASE CONTROL**

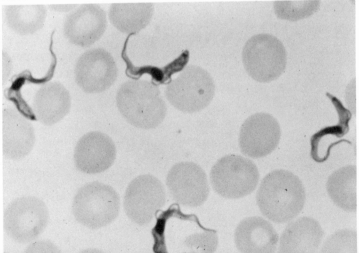

during the night. Fever attacks come and go. Weeks may go by without symptoms. Trypanosomes are in the blood in large numbers during the fever periods but are sparse when there is no fever.

Sometimes the disease does not go beyond the bouts of fever, lymph node swelling, and other symptoms. The parasites die out and that will be the end of it. If not, the disease will proceed to the next phase, in which the trypanosomes invade the nervous system. This phase does not usually start until some six months to a year after the parasites rode in on the tsetse's bite.

This final, and often fatal, phase begins slowly. The patient moves slowly, doesn't feel like doing anything, and appears to be constantly confused and, in general, unaware of what is going on. The patient starts to sleep a great deal. These periods of sleep become longer and longer, and it becomes difficult to wake the patient up. Finally, the patient goes into a coma and dies. If there has been no care for the patient, he or she will become extremely thin, to the point of emaciation. Sometimes the disease becomes chronic, not killing the patient immediately but slowly robbing him of strength. Attacks can come and go for years.

TREATMENT

The disease is treatable at all stages. Of course, the longer the disease has been in progress, the less chance there is for successful treatment. The drugs used are very potent substances and must be administered with extreme care. In the later stages of the disease, the trypanosomes are in the brain; therefore the drug has to get to the brain and kill the parasites there. The drugs can cause side effects that involve the nervous system.

Rhodesiense Trypanosomiasis

LIFE CYCLE

This disease is transmitted in the same way as gambiense, but is more severe.

SYMPTOMS

Once the patient has been bitten by a tsetse carrying *Trypanosoma brucei rhodesiense*, the course of the disease is fairly rapid. Trypanosomes are seen in the blood a few weeks after the bite. Weight loss is fast, and the patient goes into a coma and dies. Untreated patients die within nine months to a year after infection. There is less likelihood that the disease might become chronic than is the case with gambiense.

TREATMENT

Treatment is similar to that for gambiense. However, since the rhodesiense trypanosome is tougher, higher doses of drugs are usually needed, with consequent greater risk of side effects.

FACTS AND FIGURES

Slave traders operating along Africa's west coast quickly learned not to accept slaves with swollen glands in the neck. Traders and masters often beat and kicked apparently "lazy" slaves who would not move fast enough for them. No amount of beating, however, would induce these slaves to move faster, and many soon withered away, went into a coma, and died. Those with chronic sleeping sickness were regarded as lazy, and the beatings they received compounded their misery.

African sleeping sickness could well have become a problem in the Western Hemisphere if there had been a suitable insect host. It is still a serious problem in Africa.

The World Health Organization estimates that 35 million people live in areas where they are at risk of African sleeping sickness. WHO estimates that some 10,000 new cases a year occur, a figure considerably less than a hundred years ago.

Gambiense is found in scattered areas of concentration along Africa's west coast, including Gambia, Senegal, Guinea, Ghana, Nigeria, and Gabon. Much of Zaire is also gambiense sleeping sickness country. Rhodesiense is found in eastern Africa in Uganda, Tanzania, Zimbabwe-Rhodesia, Malawi, Zambia, Rwanda, and Burundi.

Rhodesiense sleeping sickness was spread into new territories by colonial exploring, trading, and land-grabbing activity. There is a possibility, for example, that the disease was introduced into Uganda by the famous Stanley expedition of 1888. A terrible epidemic followed that reduced the population is some regions to a third of what it had been before.

A prevailing pattern during the colonial era in Africa was that white settlers would push Africans out of higher, cooler land, free of tsetse flies, into low-lying areas infested with them. This movement served to spread the disease and was particularly hard on tribes in East Africa that depended on cattle-raising to make their living.

The rhodesiense form of the disease exists among wild animals. The wild animals are not bothered by the trypanosomes, but domestic cattle are. Attempts to raise cattle in tsetse country inevitably met with failure as the cattle withered and died of what the cattle-raisers called "nagana." People died, too, as the flies spread the disease from cattle to people and back to cattle, then into the wild animals and back to human settlements.

Many of the areas that are set aside as wildlife preserves in Africa are tsetse fly areas. Many wildlife conservationists find it hard to conceal a grudging admiration for the tsetse fly, an animal that in its own blind way has done more to save extinction-threatened African wildlife than all the rhetoric politicians have been able to muster. No one in his right mind will bring cattle into tsetse fly country, and in much of East Africa no cattle means no people.

Top view of a tsetse fly with a close-up of the mouth parts below. As shown here it is about three times its actual size.

THE CENTER FOR DISEASE CONTROL

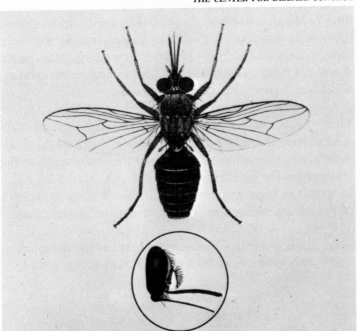

The gambiense form has no wildlife reservoirs, and therefore is easier to control than the rhodesiense form. The gambiense-spreading tsetses are generally found around rivers in the shade of riverbank vegetation. Control is achieved with insecticides and by clearing riverbanks of vegetation. The rhodesiense form is almost impossible to control. Quarantine measures are about all that can be done. East Africa tsetses live throughout the dense woodlands, rather than concentrated on riverbanks.

There are preventive medications that can be taken by injection. These are advised for those who have to travel into tsetse country.

Western Hemisphere Trypanosomiasis

Charles Darwin was sick for much of his life. He always felt tired and suffered from a great deal of abdominal distress, including pain, vomiting, and diarrhea. The doctors at the time (1840's) could not diagnose Darwin's illness. Many thought he was just a high-strung hypochondriac who only thought he was sick. His friends believed his symptoms were due to the pent-up hatred he had always had for his tyrannical father, hatred that he never released.

As it turns out, Darwin probably had a very real disease. The evidence that his disease was real and not hypochondria comes from the notes he made during the voyage of the *Beagle*, the expedition that provided him with much of the information for his theory of evolution. On March 26, 1835, he spent the night in Luxan, Chile, a small town at the foot of the Andes Mountains. He wrote that he experienced

an attack, and it deserves no less a name, of the Benchuca, the great black bug of the Pampas. It is most disgusting to feel soft, wingless insects, about an inch long, crawling over one's body; before sucking they are quite thin, but afterwards round and bloated with blood, and in this state they are easily squashed.

Darwin was sick almost from the moment he returned to England in October of 1836. There seems little doubt today that he had what is now called Chagas' disease. The cause of this disease, *Trypanosoma cruzi*, was not known until 1909, so the physicians of his time cannot be faulted for their ignorance. Had they known, it wouldn't have mattered, since we can't do much about it even today.

LIFE CYCLE

Trypanosoma cruzi look much like the trypanosomes that cause African sleeping sickness. They are generally about 20 micrometers long, but they vary widely in length. The intermediate hosts are several species of insects, among which are the "benchucas" referred to by Darwin. These insects, which are true bugs, are generally referred to as *reduviid* bugs. In older books you will see them called *triatomids*. Common names include "kissing bugs" and "barbieros." The insect that attacked Darwin was an immature nymphal form that does not have wings.

The trypanosomes live in the intestinal tract of the bugs. The form in which they occur in the bug is different from the forms they assume in human and other hosts. When the bug sucks blood, it defecates. The trypanosomes in the bug's wastes get into the host through the wound made by the bug. When a bug takes a blood meal from a host, it can take up some trypanosomes and the cycle continues.

An immature (top) nymph stage of a triatomid bug and an adult triatomid (bottom). The immature form is the one that carried out the attack Darwin described in his memoirs. When they bite, triatomids can transmit the flagellates that cause Chagas' disease.

THE CENTER FOR DISEASE CONTROL

SYMPTOMS

A sore forms where the bug has taken its blood meal. This sore can be quite large and very painful. They are usually on the face, but they can be anywhere. The sores increase rapidly in size for a few days, and then slowly go away. The sores are filled with various forms of trypanosomes. From the sore, the trypanosomes spread to the lymph nodes, which become swollen.

Some two weeks after the bite, the patient goes into what is called the acute phase. The acute phase begins with scattered swellings here and there around the body. Frequently, there is a peculiar swelling of one side of the face, particularly the eyelids. This is a sure sign of Chagas' disease. The severity of symptoms varies with the age of the patient. Infants and children are hit hardest; in adults the course of the disease is usually less severe. Infants seldom survive the disease.

During the acute phase the victim has all kinds of symptoms, including high fever, chills, muscular aches, and nosebleeds. The nosebleeds are particularly a children's symptom.

The acute phase of Chagas' disease lasts a few weeks and ends either in death, more or less complete recovery, or passage into a chronic stage. Charles Darwin probably had the disease in its chronic form. Some people who never knowingly had acute Chagas' disease have been found to have the chronic disease. The chronic disease occurs more often in adults than in children.

Over the years, the coming and going of attacks causes all kinds of damage to the internal organs. Although chronic Chagas' disease may not kill the patient outright, it does shorten and reduce the quality of life.

The heart is most frequently affected, and heart attacks are fairly common among people with chronic Chagas' disease. Other long-range effects include massive enlargements of parts of the digestive system.

FACTS AND FIGURES

Chagas' disease is found throughout South America, particularly in Brazil. It also occurs in Central America and Mexico, and there may have been some isolated cases in Texas. It is mostly a disease of the poor, who live in hovels and shacks in conditions favorable for the proliferation of the bugs.

The World Health Organization estimates for the number of people with Chagas' disease is 10 million. Some 35 million are at risk. The number of cases and the severity of the disease decreases as one travels north from South America. However, there are reduviid bugs in the United States, and trypanosomes have been found in at least fourteen wild animals common in the United States. The United States has been free from Chagas' disease for a number of reasons. House construction in the United States is not favorable to the bugs' way of life, and many species of American reduviids do not defecate as they eat. Most of the animals here who are infected with trypanosomes get them from eating the bugs and by licking fur contaminated with feces.

There are occasional cases among Americans who visit endemic areas such as Brazil.

Leishmania

Leishmania are flagellates that cause human disease in many parts of the tropical world. The flagellates are similar, but the diseases they cause have their own set of characteristics. The flagellate that causes the disease in India, for example, cannot be told apart from the one that causes a similar, but different, disease in South America or Egypt.

The intermediate hosts are insects. One of the more important is the sand fly *Phlebotomus.* Mosquitoes are also intermediate hosts for some kinds of *Leishmania.* The flagellates themselves are small, even for flagellates. Typically, the forms that exist in vertebrate hosts such as humans are oval in shape, seldom more than six micrometers long and three micrometers wide at the widest point. The forms that live in the insects are somewhat larger—around 20 micrometers long at the most and 3 micrometers wide.

LIFE CYCLE

In the insect host, the flagellates develop in the gut. The infective forms of the parasite move forward to the mouth parts of the insect. There can be so many of them that the mouth parts of the insect are blocked. When the insect tries to get a blood meal, the mass of flagellates is often broken up and some of them then get into the wound caused by the bite. Not as elegant a scheme as the neat saliva squirts of anopheline mosquitoes, but effective nonetheless.

Once *Leishmania* invade man or some other vertebrate host, they reproduce in white blood cells and other cells and tissues. The newly formed flagellates break out of the cells to invade more cells. Many of the flagellates, however, are ingested by white blood cells.

There are two or three species, depending on which classification you follow, that are major causes of disease in man. For the sake of simplicity, we will assume there are three:

Leishmania donovani, *Leishmania tropica*, and *Leishmania braziliensis*. The latter two are considered variations of the same species by some parasitologists.

Leishmania donovani

The disease caused by this flagellate goes by many different names around the world. It is most commonly called kala-azar. In some places it is called dumdum fever and black disease. Physicians refer to it as visceral leishmaniasis or tropical splenomegaly. In some parts of the world dogs are as important a host as man. The intermediate host is the sand fly, although there is some evidence that direct reinfection can occur. In India, kala-azar is mainly a disease of adults, and dogs do not get it at all. In the countries along the Mediterranean Sea, it is primarily a disease of children, so much so that it is generally called infantile kala-azar. The Mediterranean strain of the disease does occur in dogs and is also prevalent in Central and South America, China, mid-Asia, the Sudan Republic, and in Kenya. Rats are reservoir hosts in the Sudan, and ground squirrels and gerbils are reservoir hosts in Kenya.

SYMPTOMS

Kala-azar is a debilitating disease with many different symptoms, such as diarrhea. Other symptoms are malaria-like. The disease is called visceral leishmaniasis because most of the activity of the flagellates is in the internal organs rather than near the surface of the skin as in the other two species. The spleen and liver are frequent sites of concentrations of the flagellates. The spleen is often grossly enlarged, hence the name "splenomegaly."

As is the case with many parasitic diseases, kala-azar is not an outright killer, but it does rob its victims of energy and drive. In so doing, it affects the overall economy of the regions in which it is prevalent.

TREATMENT

Kala-azar can be treated with drugs made of antimony compounds. For years the treatment of choice was trivalent potassium antimony tartrate or pentavalent antimonials. The latter are usually less toxic and more effective than the trivalent compounds. A disadvantage of both types of antimony compounds is that in a given patient, they tend to become less effective with use. In recent years a group of drugs called the diamidine series has been shown to be effective. These drugs cause fewer side effects in most patients than the antimonials. Splenectomy (removal of the spleen), followed by intensive drug therapy, is a very effective treatment, often resulting in a complete cure.

Leishmania tropica

Common names for this disease include oriental sore, Delhi ulcer, Aleppo, and Baghdad boil. Doctors call it cutaneous leishmaniasis. The word cutaneous refers to the skin area, and that is where this parasite does a great deal of its damage.

The intermediate host is *Phlebotomus*, and the life cycle is about the same as that of *L. donovani.* Gerbils and other wild rodents and dogs and cats are reservoir hosts. The names given to the disease suggest where it occurs: the Middle East, mid- and southwest Asia, North Africa, and Central and South America.

SYMPTOMS

The major symptoms are skin ulcers. Some time after the insect bites, a small red pimple appears at the bite site. The time between the bite and the appearance of the pimple (the incubation period) can be as little as two weeks or as long as three years. Two to six months is the average. The pimple increases in size, itching as it grows. The skin breaks and an ulcer forms.

Two types of ulcers, wet and dry, are recognized. The wet, or moist, type occurs in rural areas. The ulcers tend to form quickly after the bite, run a rather acute course, and then fade away. This type is usually transmitted from gerbil to man via the sand fly, and is more common in rural areas because gerbils are more likely to be in the country than in the city. The dry or city type is much slower in erupting after the bite. It tends to last longer and often becomes chronic. Usually, however, the sores of both types heal, leaving scars that lack pigmentation and are slightly depressed.

TREATMENT

One attack generally confers a degree of immunity. This fact is widely known in endemic areas, and some people "vaccinate" children by allowing a fly to bite in a part of the body where a scar will not be disfiguring. There are treatments, but it is often better to forget about treatment in endemic areas if the ulcer is not on the face or any other place where the scar will be disfiguring. If the patient does not live in an endemic area but picked it up while traveling through, treatment is generally advisable. A drug called sodium antimony gluconate has been found to be effective. The drug is very toxic and must be used with great care. Plastic surgery is sometimes needed to repair particularly disfiguring scars. The sores are almost always infected by bacteria of one kind or another. These infections are treated with antibiotics.

Leishmania braziliense

This flagellate can, in certain conditions, cause one of the most utterly disgusting, disfiguring diseases ever to afflict mankind. Depending on the nature of the disease and the part of the world in which it occurs, the disease is called espundia, chiclero ulcer, uta, or forest yaws. Medical terms include mucocutaneous leishmaniasis, American leishmaniasis, and nasopharyngeal leishmaniasis.

SYMPTOMS

Various types of the disease are recognized. In Mexico, particularly in the Yucatan area, the type called chiclero ulcer is prevalent. The name comes from the *chicleros*, men who go into the Yucatan jungle to collect chicle, which is used in the manufacture of chewing gum. Chicleros have a reputation as a rough-and-tough bunch of guys from whom it is best to stay away. The sand flies, however, don't stay away and the chicleros are often bitten. Typically, the ulcers appear on the ears in the chiclero type of leishmaniasis. The disease is generally chronic and does not seem to keep the chicleros from going about their business. Why the ulcers are mostly on the ears is not known. However, it is thought that the sand flies in the jungle areas where the chicleros work bear a particular strain of *Leishmania* that has an affinity for the ears.

The uta type is much the same as oriental sore. This type is found in higher elevations in Mexico.

Espundia is potentially the most spectacular and destructive form of the disease. It occurs mostly in rural areas of Brazil, generally in lower elevations that tend to be hot and humid. In espundia, the ulcers are massive; frequently, several ulcers blend into a single festering mass. Unfortunately, these proliferating ulcers occur most frequently on the face. In extreme cases, the nose and parts of the mouth, such as the palate, can be completely destroyed. The effect is horribly disfiguring. Bacterial infections make the situation a great deal worse. Death, ordinarily from bacterial infection, is not infrequent. The severe form of the disease occurs mostly among people who work in the deep forest in such occupations as lumbering.

TREATMENT

The disease can be treated. However, early diagnosis is essential so that treatment can begin before damage to the tissues goes too far. The same drugs that are used for other forms of leishmaniasis are also used for espundia. Plastic surgery can sometimes repair the damage done by the flagellates.

Giardia lamblia

Giardia lamblia is a pear-shaped flagellate that appears to have a sort of "monkey face" if viewed from the right angle with a microscope. Of course, the organism has no face at all. What appear to be the eyes of the "monkey" are cell nuclei. G. lamblia live in the intestinal tract. They are thought to eat semidigested food there, but are also known to cling to the intestinal mucosa, so they might indulge in a bit of intestinal tissue from time to time.

Giardia is a fairly common parasite all over the world, including the United States. It is spread in food contaminated by sewage in a hand-to-mouth manner. There have been epidemics in some very posh places, one of the most famous of which was in the ski resort town of Aspen, Colorado.

The symptoms generally range from none to bothersome to severe. The parasite hits children harder than it does adults. Children can suffer from irritation and inflammation of the intestinal mucosa, and diarrhea alternating with constipation. Frequently, what parents might think of as colic is a bit of *Giardia.*

Treatment with quinacrine is usually very effective, although reinfection is quite common.

Trichomonas **Spp.**

Trichomonas is a genus of flagellates that is widely distributed in various parts of many animals, particularly domestic animals. Three species are known to inhabit human animals at one time or another. *T. hominis* lives in the intestine, *T. tenax* is an occasional tenant of the mouth, and *T. vaginalis* is a frequent visitor of the female reproductive tract and an occasional invader of the male reproduction system.

T. vaginalis is a widespread human parasite. It lives in the vagina of the female, where it can, in certain conditions, cause an unpleasant, frequently chronic inflammation of the vagina. Patients complain of itching and burning sensations, and there is often a discharge from the vagina associated with the disease. It is also found in the male reproductive system. No one is quite sure how frequently it inhabits the male anatomy, since it is more difficult to find them there than in the vagina, and they seldom cause noteworthy symptoms in males.

The life cycle is a bit of a mystery. *T. vaginalis* is probably spread by sexual relations, but very young girls, even infants, have had the infection, so there are other means of spreading the parasite.

Many effective treatments are available. These treatments are the usual kinds of medications taken in pill form, and medications applied directly to the vagina. Metroxidiazide is a frequently used pill-form medication. Vagisec is the trade name of one of the more commonly used medications applied directly to the vagina. The incidence of trichomonas may be as high as 25 percent, but not all infections cause symptoms. The key seems to be the pH of the vagina (acidity and alkalinity). The normal pH of the vagina is 3.8 to 4.4. *Trichomonas* cannot live at that pH: The pH must be above 4.9 in order for *Trichomonas* to survive. Hence, it would seem that *Trichomonas* is an opportunist that takes advantage of unhealthy conditions, which can be brought on by any number of factors. One of the factors that has contributed to an increase in the incidence of *Trichomonas* in recent years is panty hose. Panty hose reduce ventilation of the vaginal area, a circumstance that tends to raise vaginal pH.

8.

Arthropods

Suppose some scientist from another planet came here in his space ship to survey the kinds of life that exist on this world. The report of this expedition could well be titled "A Planet of Arthropods."

We like to think that we are the dominant form of life on this planet, and in many respects we are. In sheer numbers we are probably the most numerous single species of animal. As a group, however, our numbers are puny indeed compared to the arthropods. There are more species of arthropods on this planet than there are all other species combined.

The word arthropod means "jointed leg." They do indeed have jointed legs, and they also have their skeletons on the outside of the body rather than on the inside. Arthropods include animals such as crabs, lobster, crayfish, and shrimp. These make up a class of arthropods called Crustacea. Another class of arthropods are the Arachnida. These include spiders, scorpions, mites, and ticks. Centipedes and millipedes are members of separate classes.

By far the largest class of arthropods are the Insecta—the insects. There are about five times as many species of insects as there are all other species of animals combined.

Insects and humankind have long been in competition for the same needs. Constant effort is required to keep insects from eating all our food and fiber crops. Although they are in competition with us, very few insects or arthropods obtain their life needs through parasitism. The few that are parasites tend to be only occasionally so, and almost all of those are ectoparasites. As far as humanity is concerned, only two classes of arthropods—insects and arachnids—include parasitic species of any interest.

Insects

Insects are those arthropods that have six legs and three distinct body parts. They are found everywhere. We have already had a few things to say about some insects with more or less parasitic habits. Mosquitoes, black flies, sand flies, and tsetse flies can all be considered to be parasites at some times in their lives by virtue of their need to suck blood. Of greater importance than the little bit of blood they take is the fact that harmful parasites of many varieties can be introduced into the host while they sup.

There are, however, some insects whose parasitic activities are of a more permanent nature than occasional blood-hunting expeditions. Among these more companionate inhabitants of our bodies are lice. The fact that they can introduce other unwanted inhabitants into our interiors compounds the problem.

Lice

"Lousy" is one of the more expressive words in our language. Most of the time the word is used in a context far removed from its original meaning, which was "infested with lice." Today lousy means no good, terrible, and so on. In recent years however, the word has been used in its original context with increasing frequency. Human lice infestations are on the increase, so much so that many public health officials are using the word "epidemic."

A louse is a small animal, even for an insect, but they can be seen with the naked eye, a circumstance that is often the cause of much distress. They are dark, rather flat insects. The female is about 3 millimeters long and the male is about 2 millimeters long. There is no intermediate host; the human animal is the only one it ever needs.

There are three kinds of human lice: head, body, and pubic. The head and body varieties are identical in appearance; the only way to tell them apart is by where they live. The head louse goes by the name of *Pediculus humanus capitis*, and its body-dwelling cousin is called *Pediculus humanus humanus*. The pubic louse differs distinctly in appearance from body and head lice. Its name is *Phtirus pubis*.

PEDICULUS HUMANUS CAPITIS

Head lice live on the head among the hair. The females lay eggs on the individual hairs. The eggs are firmly glued to the hair. Referred to as "nits," the eggs are quite visible as white, shiny oval structures that combing will not remove. The eggs hatch about ten days after they are laid. What emerges from the eggs are nymphs, similar to but not quite like the adults. They mature in about two weeks. Many lice stay on the same host on which they were born for their entire eighteen-day lives. Others venture to new heads. The little insects are helped in their travels when people come close to each other.

Human head louse, Pediculus humanus capitis.

PEDICULUS HUMANUS HUMANUS

The body louse is similar to the head louse in every respect except that it lives on the body rather than the head. Actually the lice spend more time on clothing than on the body that is covered by the clothing. The lice wander to the body surface from time to time for a bit of food. When clothing is removed, most of the lice remain in the clothing rather than on the body. This indicates that body lice evolved after man began wearing clothes. Body lice are spread by close contact with the host or by contact with infested clothing. Sometimes they can be picked up from sitting on fabric-covered seats in theaters and other public places.

Lice are easy enough to get rid of, but they have a way of coming back, especially if you are in an area where they are abundant. Many medications are available. Benzene hydrochloride is one regarded as effective. For head lice, the hair and scalp are washed thoroughly, rinsed, and damp-dried. The medication is applied and worked into the scalp, then left there overnight. A repeat treatment a week later is usually recommended to take care of any survivors of the first killing. For body lice the medication is applied in powdered form to the body and clothing. All affected clothing should be boiled or discarded.

Lice are troublesome in a number of ways. Knowing that you have them is a bit embarrasing. Their bites can be irritating, and some people react to the bites more severely than others. The lice secrete a little saliva when they bite, and that can cause an allergic reaction. Itchy red bumps rise at the bite areas. The itching encourages scratching, which encourages secondary infection from bacteria.

Another way lice can cause distress is by carrying disease organisms. The body louse is known to carry typhus, a serious, often fatal, disease. The lice pick up the disease organisms when they bite an infected person. New hosts are infected when the organisms get into the body through the bite wounds. Other diseases carried by lice are trench fever and relapsing fever.

Lice and typhus epidemics often occur during and immediately after wars and other upheavals. In battle situations soldiers may be in the field for long periods of time without much opportunity to wash or change clothes. Lice thrive in situations such as these, and so does typhus. Typhus causes high fever, and lice tend to leave bodies with abnormally high temperatures to seek cooler ones, which serves to spread the infection. The civilian population also suffers. There were massive epidemics of lice in Europe during and after World War II and during the Korean War. Typhus was a particular problem during the Korean War.

Lice, also called cooties and motorized dandruff, among other things, are not limited to wartime, nor are they limited to "dirty" people. Throughout the 1970's, there was a steady increase in the number of people with lice in the United States. Companies that make anti-lice medications could hardly keep up with the demand for their products.

Most of the victims of the epidemic are children and teen-agers. The long hairstyles that became popular in the 1960's may be one reason for the epidemic, but that is not the complete answer. School is as good a place as any to pick up lice, but that is no reason not to go to school. Lice can be controlled with frequent hair washings and not wearing the same clothes too long between washings or cleanings. During large-scale epidemics such as those that occur during wars, mass dusting of people with insecticide powder is an effective control.

PHTIRUS PUBIS

Phtirus pubis is also on the increase. This parasite, commonly known as crabs or blue crabs, lives mostly in the pubic hair, although it can be domiciled elsewhere in other hairy parts of the body. They are most often spread by sexual intercourse, but they can be picked up from toilet seats, towels, and other contaminated objects. They are not known to transmit any disease, but their presence can be uncomfortable and disconcerting. They have short, stubby legs

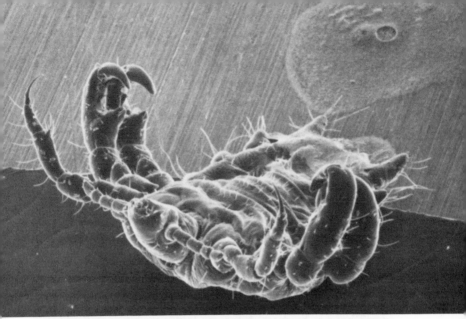

(Top) a scanning electron micrograph of Phtirus pubis, *the human crab louse. (Bottom) an unknown Japanese artist's interpretation of the same organism.*   **AMERICAN METALS RESEARCH**

equipped with claws that are used to hold on tightly to the pubic hairs. The same drugs that are used for head and body lice are also used for pubic lice.

The increase in the number of crab lice cases is thought to be due to the increase in sexual activity among teenagers. They sometimes cause bluish spots on the skin, hence the name "blue crabs."

Fleas

Fleas are a frequent problem of dogs, cats, and other mammals and birds. They sometimes afflict people too, and you don't have to be around dogs and cats to pick up a few every now and then. Fleas are bloodsucking insects. They stay on the host for varying periods of time, depending on the species, while they get their blood meals. Some can stay alive between feedings for as long as four months. Others would perish in a month without something to eat.

An occasional flea bite from an errant dog or cat flea is usually of no great consequence. However, large numbers of bites can result in skin irritations. The main trouble comes from some tropical species and from species that can carry disease such as bubonic plague and typhus. Also, some dog fleas are intermediate hosts of the most common tapeworm of dogs, a tapeworm that can invade people.

The most damaging tropical flea is called *Tunga penetrans.* The second part of the name gives an idea of what this flea can do. The flea goes by a number of names in Mexico and Central and South America, where it is chiefly found (it also exists in some parts of Africa). Chigoe, nigua sand flea, and jigger are several of the names given to this annoying insect.

The female is the one that does the damage. She burrows into the skin to deposit her eggs. As she fills with eggs, she balloons to an impressive 2–3 millimeters in diameter. Often she burrows between the toes and under the toenails, her rear end protruding so the eggs can drop out. The frequent result is a large, festering, painful ulcer. The only way to get the female out is with surgery. Sometimes secondary infection occurs. Gangrene, occasionally bad enough to result in loss of a limb, can be the result.

The lowly flea, particularly the rat flea *Xenopsylla cheopis*, and other fleas have figured mightily in human history. The flea can carry bubonic plague from rat to man. Called the Black Plague and the Black Death, the disease killed more than one-fourth of the population of Europe in the fourteenth century. There were other epidemics in the 1640's and 1650's, and cases appeared in the United States along the West Coast in the 1900's.

Plague is caused by a bacterium called *Pasturella pestis*. It is mainly a disease of rats and other rodents. From time to time, however, usually through human activity, plague overflows from its natural base among rats and into the human population.

When rats die of plague, the fleas leave the body of the dead animal as it gets cold. The fleas are attracted to a new warm host, and if most of the rats around are dead, the new host is likely to be human. Typically plague has broken out among rats in seaport cities when infected rats came ashore from ships. The disease quickly spreads through the local rats, killing so many of them that the fleas start to jump on people.

Fleas that have left a dead rat are loaded with *Pasturella pestis*. When they bite a human host, the bacteria can get into the host through the bite, sometimes from contamination by the flea's feces. The flea might have so many bacteria in its gut that it is blocked, and attempts to take a blood meal can result in the regurgitation of the bacteria into the bite wound.

Two to four days after being bitten, the victim develops a high fever and the lymph nodes swell up and become bulges called "buboes," the characteristic symptom of the phase of the disease called bubonic plague. Another form of the disease that develops during epidemics is septicemic plague, in which the bacteria invade the bloodstream in large numbers. The bacteria then get into the lungs and the disease assumes its most dangerous form—pneumonic plague. Pneumonic plague is spread through the air in a direct person-to-person way. Bubonic plague can generally be cured with antibiotics, septicemic plague is less likely of cure, and pneumonic plague is almost always fatal.

Plague is thought to have been brought to Europe through the 1340's campaigns of the Tartar prince Janiberg Khan. The people he ruled were the descendants of the Golden Horde that had come from Mongolia a century earlier. His army campaigned through an area, now part of the southern Soviet Union, called the Kirghiz Steppes. That area is still a breeding ground for all kinds of wild rodents. The soldiers camped in the field, came into close contact with the rodents, and many of the men picked up the plague.

Janiberg Khan became allied with Venice in a war against Genoa. At the time the Italian city-states were always fighting wars with each other. The Tartar army laid siege to the town of Kaffa, which at the time belonged to Genoa, in the Russian Crimea.

During the siege the Tartar soldiers started to die by the hundreds. According to one account, ". . . the humors coagulated in the groins and they developed a subsequent putrid fever and died. . . ." The Tartars loaded dead soldiers into

Protective clothing worn by a physician during a plague epidemic in Italy in 1656. The drawing appeared in a book published in 1661.

their catapults and hurled them over the city walls. Inhabitants of Kaffa started to die of plague in a few days. Although the people involved knew nothing about bacteria, this incident may have been the first use of bacteriological warfare. Citizens of Kaffa fled in terror. They took ships to ports all over the Mediterranean Sea, carrying the plague with them; the Black Plague of the fourteenth century had begun.

Conditions in Europe at the time contributed to the spread of the plague. There were frequent wars during this century, resulting in almost constant movement of armies and people fleeing the strife. People lived crowded together in cities in close communication with rats and therefore with fleas.

Outbreaks today tend to be short and sharp. The fleas do

not keep the infection indefinitely, particularly in warm regions, and many fleas die of the infection. These circumstances and the use of antibiotics help to bring outbreaks to a close before there is much chance of development of the pneumonic form of the disease. However, the infection may not go away entirely.

Plague exists in a form called sylvatic plague along the edges of many large cities. On the edges of some cities, city-type rats intermingle with wild rodents such as ground squirrels. If a plague epidemic has occurred in the city, some of the wild rats will become infected from the city rats. The infection can then exist among the wild population for many years.

One such area of sylvatic plague exists outside of San Francisco. Ordinarily the infection can be expected to remain among the wild rodents, but if a person handles a rodent that has died of the plague he or she can pick up the disease and possibly pass it on through the pneumonic form. This has happened in and around San Francisco a few times since the first outbreaks there in the 1900's. The danger of this sort of thing happening increases as the suburbs push farther out into the countryside. We may never see a Black Death in this country, but isolated cases of plague will continue to occur.

Bedbugs

Bedbugs or "cheenchas," as they are sometimes called in the American South, are not as big a problem as they used to be. Their decline as a major irritant of humanity has been brought about mainly through wider use of metal bedsteads rather than wooden ones. Typically these brownish, oval, 5.5-millimeter-long insects hide by day in cracks and crevices of wooden bedsteads, under wallpaper, in baseboards, and other such places. At night the bugs come out of their hiding places for a meal. Their fare is always blood, but they don't eat much—they can go for a year without eating.

People react with differing degrees of severity to the bites. Some who are particularly allergic develop itchy red sores that may become infected. Others show no reaction at all. From time to time a parasitologist advances the idea they might carry some disease, but no one has been able to prove this accusation.

Flies

Flies such as tsetses that are vectors of parasites have been discussed in previous chapters. Some kinds of flies can cause a problem called myiasis, which is maggot infestation. The maggots come from eggs laid in wounds by various species of flies. Myiasis is an infrequent, but not too rare occurrence in man. People most often affected by it are sheep and cattle raisers. Sheep and cattle, on the other hand, are plagued by a number of flies that deposit their eggs in open wounds and in the noses and eyes of these animals.

Although a myiasis can be a painful and unpleasant experience, it is usually not as bad as it sounds and is readily treatable. In World War I maggots were used medically to clear away gangrenous flesh from infected wounds. Wounds so treated healed more rapidly than those covered by bandages.

The common housefly, by the way, does not cause myiasis. However, our familiar friend has been implicated in spreading *Giardia* and other disease-causing organisms through phoresis.

Arachnids

Spiders, ticks, and mites are commonly thought of as insects by many people who would prefer to think as little as possible of spiders, ticks, or any other such animal. The arachnids are arthropods that have eight legs as adults and no wings. Like insects, most are free-living. Those who are parasitic are mostly of the occasional variety, and are ectoparasites. There are a few that are endoparasites in their own peculiar way.

Ticks

Ticks are bloodsuckers, and there are few mammals, birds, reptiles, and amphibians that escape their attention. Practically any of them could take a blood meal from a human host if given the opportunity. Many ticks get that opportunity when people go hiking or camping in wooded areas.

There are two basic kinds of ticks, hard and soft; the hard are the ones most likely to attack a human host. A genus called *Dermacentor* is the most common visitor to our skin surface. It is reddish brown, and varies greatly in size depending on how much blood it has ingested. Both the male and female are bloodsuckers. They pierce the skin with sharp mouth parts and drive an effective "drinking straw" into the skin. Once they pierce the skin they remain firmly attached for several days while they gorge themselves on blood. The time spent in blood-feasting varies from five to fourteen days.

After she has had her fill, the female drops off and lays eggs in the ground. In a matter of weeks, she lays anywhere from 2,000 to 8,000 eggs and then dies. The baby or seed ticks attach themselves to small animals for a drink, drop off, and then mature into a nymph stage. Generally they hibernate over the winter, take another blood meal or two when the weather gets warmer, and then mature into adults on the ground. When matured, they wait for an opportunity to find a new host.

Tick bites can be painful and bothersome. The holes they make in the skin are more than big enough to encourage secondary infeçtion, sometimes opening the way for myiasis in domestic animals.

Ticks can cause a serious condition called tick paralysis. A toxic substance given off by the ticks interferes with the normal chemistry of the nervous system. The paralysis spreads gradually, and there can be permanent damage to the nervous system. This paralysis occurs most often in domestic animals, but it can occur in people. The paralysis stops if the ticks are removed.

Tick. Amblyomma americanum *is commonly called the lone star* *tick.*   **THE CENTER FOR DISEASE CONTROL**

Ticks have been implicated in the transmission of some pretty awful diseases. The list is a long one indeed and includes Rocky Mountain spotted fever, Q fever, hemorrhagic fevers, relapsing fever, Lyme arthritis, and tularemia. In 1977 there was an outbreak of babesiasis on the island of Martha's Vineyard off the coast of Massachusetts. Caused by a sporozoan, the disease is carried by deer ticks. All of these diseases are serious matters, and some of them are incurable.

Ticks should be removed as soon as they are seen, but it must be done carefully. Just pulling them out will usually leave the mouth and head parts in the wound, which might cause an infection. The best way to remove the tick is to apply something that will weaken it first. Among the substances that can be so used are alcohol, gasoline, kerosene, and ethyl chloride. Another trick is to hold a lighted match or cigarette close to the tick. Needless to say, one does not use the lighted match or cigarette method if any of the mentioned liquids have been applied to the skin!

Often the solvents or the heat will cause the tick to back out on its own. If not, pull it out carefully. This is best done with a pair of tweezers. Clasp the tweezers as close down to the very front of the tick as you can. Pull gently and carefully, taking care not to break the mouth and head parts. Always wash your hands thoroughly after removing ticks.

Ticks are tough characters, difficult to kill. They can survive repeated stompings from the heaviest boot-clad foot. The best thing to do with removed ticks is to put them in a bottle of alcohol. Cover the bottle and leave them there for several days.

Mites

Mites are like ticks, only smaller. Most are free-living. Quite a few live among decaying vegetation and some cause economic loss by contaminating stored food. The parasitic forms are widely distributed among many kinds of animals and plants. Some cause plant diseases that take much of the joy out of growing houseplants.

Most mites are so small they can barely be seen. But they can easily be seen with a microscope, and looking at different species of mites thus enlarged will reward you with sights of some of the most weird, bizarre, and outlandish-looking creatures you will ever see.

Chiggers

Summer camp is a pleasant experience for some and not so pleasant for others, but nothing, not the meanest counseler or the most annoying bunkhouse mate, can take the joy out of camping more effectively than chiggers can. *Trombicula alfreddugesi*, also called red bugs, are tiny red mites that feed on the blood of a variety of animals including man. The mites abound in bushes, grass, and other vegetation, where they wait to get on any passing source of a blood meal. Typically they crawl up the legs and attach themselves firmly to the skin. Often they concentrate under the tops of socks.

Their bites cause an intense itching and swelling that reaches a maximum of discomfort in two days. After the swelling goes down, the surrounding skin turns red and stays that way for a while.

Large numbers of chiggers can bring on a fever, especially in young children. Scratching, as always, can promote a secondary infection. Relief can be found by applying soothing ointments and taking antihistamines to reduce the swelling.

Sarcoptes scabei

Among all the thousands of organisms that can cause human disease, the itch mite, *Sarcoptes scabei*, has a special distinction. It was the first organism to be positively identified as the cause of a specific human disease. It was so designated in 1687 by an Italian physician, Giovanni Cosimo Bonomo.

The itch mite is one of the smaller mites. The female measures from 330 to 450 micrometers, and the male attains a length of no more than 250 micrometers. They burrow into the skin of their host, and, for their size, they burrow fairly deeply. As soon as the female makes contact with a host, she gets right down to business. She wanders over the surface of the skin at a rapid pace for so small a creature. Her speed has been clocked at 2.5 centimeters (about an inch) a minute. That may not sound very fast, but that distance is about 550 times the body length of the mite. For a six-foot-tall human being, that would be roughly the equivalent of running a mile in a minute and forty seconds. When she finds what to her is a suitable place, she starts to burrow a tunnel into the skin. The burrowing is a bit slower than her walking, but still impressive: two to three millimeters a day. The males dig a few side tunnels that connect with the main channel dug by the female.

Once the female has dug her tunnel, she deposits eggs for the rest of her five-week life. The usual number is from 40 to 50 eggs per burrow. The eggs are enormous compared to the size of their mother; they are almost half the size of the female mite. The larvae emerge in three to ten days. The young ones dig lateral branches or burrow new main tunnels.

Obviously all this activity does the skin no good, either in appearance or comfort. The skin rises in reddish swellings that itch unbearably. Scratching makes the situation worse through irritation and secondary infection. In heavy infections, there is a very noticeable red rash, characterized by fairly sizable, slightly reddish spots. The disease is called scabies in humans, mange in dogs.

The hands, particularly between the fingers, are frequent sites of infection. Any part of the body can be affected, however. The mites are spread from person to person by close contact.

Scabies was thought to be a relatively rare condition in the United States, but in recent years, the number of cases has gone up sharply, so much so that dermatologists were talking about a scabies pandemic in the late 1970's. No one knows why there has been an increase. Part of the reason may be that it is being diagnosed with greater frequency. Scabies is

Sarcoptes, *the human itch mite.* THE CENTER FOR DISEASE CONTROL

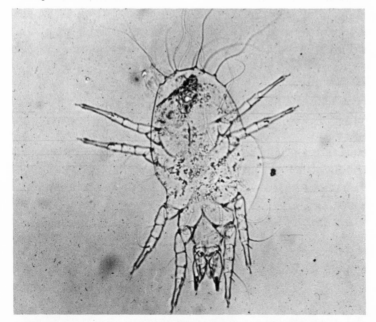

a great imitator, mimicking the symptoms of many other skin conditions.

The scabies that were seen most often during the 1979 and 1980 pandemic were not the typical variety but certain special forms of the disease. One type was particularly prevalent among kidney transplant patients who were receiving therapy to cut down on the chances of the transplanted organ being rejected by the body.

A high incidence of scabies occurred among children recently arrived in the United States from Korea and Vietnam. Also on the increase is a type of scabies transmitted to people from dogs. The dog itch mite is not quite the same as the human itch mite. Nevertheless, the dog mite can cause some discomfort on human hosts. Fortunately, dog scabies mites do not usually last very long on human skin.

Scabies is usually treatable with medications applied directly to the skin and with oral medications designed to relieve the swelling. Chances of reinfection are cut down by frequent changing of bed linens and underwear. Using too much of the medications can sometimes result in a skin irritation as bothersome as the one caused by the parasite.

Other mites

Sometimes, people pick up mites that more properly belong on various animals. For example, poultry workers can be infested by the red chicken mite. The color refers to the mite, not the chicken. They are bloodsucking mites and, although they remain on a human host only a day or two, they can make their presence amply known in that short time with their very annoying bites.

Another group of mites referred to as the *Tyroglyphidae*

cause a condition called grocers' itch in many places. The mites generally live in stored grain, flour, sugar, and other more or less dry foods. When the food is handled, the mites sometimes get on the people working with the food. The disease is less likely to occur now that machines do much of the food packaging and bins and barrels in grocery stores have been replaced by tightly wrapped packages on supermarket shelves. Packaged foods can contain dead or living mites, though.

Some mites, natural parasites of various animals, including seals, monkeys, and snakes, live in lungs, intestines, the urinary bladder, and other internal parts. In humans, however, mites living internally are likely to be accidental guests. They cause a variety of symptoms, such as asthma, coughing, and general irritation. Occasionally mites can get into the human intestine. A type of German cheese called *Milbenkäse* gets its flavor from mites that are deliberately put into it. Eating a piece of this cheese means eating several hundred mites along with their waste products. Those who indulge in this gastronomic adventure for the first time might suffer from what amounts to a colossal bellyache, but one is supposed to get used to it as one develops a taste for the cheese.

Occasionally, mites are found thriving in human urinary bladders. How they get there and what they are doing there remains a mystery. Apparently they do no harm to the host in this strange abode.

9.
Man's Best Friend?

Dogs and cats have been on companionate terms with humanity for a long time; dogs probably for at least forty thousand years, cats for a shorter time. Forty thousand years may seem like a long time, but in evolutionary time it is but a blink of an eye. It is not enough time for parasites to have evolved extremely advantageous routes of infection between man and his animals, but there may be some exceptions. The human *Ascaris* may be a form of pig *Ascaris* adapted to life in the human intestine. People can get a variety of parasites from their pets and domestic animals. However, this transmission of parasites is almost always due to carelessness and thoughtlessness of people, not to any marvelous adaptations of the parasite. In this chapter, we will be concerned with some of the parasites we might haphazardly acquire from pet animals, and with some of the common parasites of pets and domestic animals.

Parasites Humans Can Get from Pets
Dipylidium caninum

A running joke in the comic strip *Peanuts* is Lucy screaming something to the effect of, "Yecch, dog germs!" whenever

Snoopy gives her a kiss. Lucy may not be right when she tells Linus that fire hydrants grow in the ground, but she does have a point with her cry of "dog germs."

The dog tapeworm was mentioned in Chapter 2 as an example of a parasite that has evolved into the use of an intermediate host closely associated with the main host. The dog gets the tapeworm when it swallows a flea after being bitten by it. Quite often infected fleas are torn apart by the dog, with the result that larval tapeworms accumulate in the dog's mouth. If a person is licked on the mouth by that dog, the larval tapeworms can get into the human mouth and be swallowed. Infections can also occur from licked hands brought to mouths.

Most cases of *Dipylidium* occur among young children who play too intimately with dogs. Fortunately, the worms seldom cause any symptoms or do any harm to their human hosts. The worms can be kept out of both dog and master by keeping the dog free of fleas and having the dog wormed at least once a year. Treatment of human *Dipylidium* infections is usually easy. Since the worm is not in its usual environment, its hold is not very tenacious.

Visceral larva migrans

Dogs and cats commonly harbor roundworms similar to *Ascaris*, but considerably smaller than the human roundworm. These worms are *Toxocara canis* in dogs and *Toxocara cati* in cats. Ordinarily, these worms do not become established in human intestines, but worms that hatch there from accidentally ingested eggs can cause a heap of trouble in the trying.

These worms exhibit a fairly complex, still not completely

understood, development in dogs. In puppies, eggs hatch in
the intestine and the larval worms burrow into the intestine
and go on a trip similar to that of *Ascaris*, finally settling in
the intestine. In older dogs, the worms do not complete
development, but wander about in the tissues until they become encysted. In a pregnant female, however, the larval
worms can break through the placenta and get into the developing puppies, where they mature into adult intestinal
worms. All puppies have to be wormed after they are a few
weeks old. Almost all are born with the worms, and those
that aren't pick them up in the first few weeks of life. In cats,
the situation is different. The worms develop to maturity in
the feline intestine without going on any trips.

Eggs of Toxocara cati, *the common ascarid of cats. A newly
hatched larva can be seen emerging from one of the eggs.*
YALE UNIVERSITY SCHOOL OF MEDICINE,
DEPARTMENT OF COMPARATIVE MEDICINE.
PHOTO BY ROBERT SULLIVAN

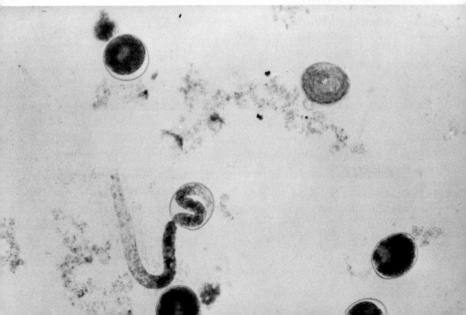

If eggs of *T. canis* or *T. cati* are ingested by man, the development will be similar to that in adult dogs. The worms hatch out and wander about in the tissues. The disease caused by the wanderings of these dog-and-cat worms can range from the inconsequential to the fatal. Symptoms include enlarged liver, lung inflammation, coughing, and fever. The larval worms can get into the retina of the eye, causing vision difficulties. In the brain, they can bring on symptoms similar to epilepsy.

The worms can bring on attacks of poliomyelitis. Polio virus is ordinarily an inhabitant of the human intestine. As long as it stays there, it does no harm. Only when it gets to the central nervous system does the virus cause the paralytic symptoms of polio. There is some suspicion that the worms carry polio virus to the central nervous system. Of course, if the host was immunized against polio, there is no danger of this complication.

There is treatment, but the disease usually takes care of itself. The worms cannot live long in the wrong host. Treatment is usually to relieve symptoms rather than kill the worms.

Again, as is usually the case in diseases of this kind, the victims are often children. Children play in the dirt and are more likely to be exposed to worm eggs deposited there by dogs and cats. Dog licking is also a mode of infection. An important preventive measure is the covering of children's sandboxes when they are not in use.

Creeping Eruption

Dogs are parasitized by two varieties of hookworms, *Ancylosotoma caninum* and *Ancylosotoma braziliense.* These hookworms cannot get to the human intestine, but they can cause problems trying to get there. The life cycles of the dog hookworms are similar in every respect to that of human hookworms, except that the usual host is a dog. However, the larval worms can enter human skin, and when they do so, they will wander around under the skin for some time. They become trapped in the tissues, surviving for weeks and sometimes months.

The wandering worms can cause the legs to break out in unsightly itching, painful lesions called creeping eruption. The legs appear to be covered with tortuous twisted tattoos. Sometimes these lesions become crusted.

Creeping eruption is why dogs are not allowed on public beaches. People are likely to be without shoes on the beach, and any dog that is on the beach will probably defecate there. Regular worming of dogs and strict exclusion of our canine friends from beaches should keep creeping eruption a rare condition.

Toxoplasma **Spp.**

Toxoplasma is a sporozoan, similar to the parasites that cause malaria. In some situations, *Toxoplasma* is the most serious parasitic infection people can obtain from association with animals, particularly cats.

As is the case with all sporozoans, *Toxoplasma* has a complicated life cycle. There are still some details of the life cycle that are not completely understood. However, it is known that people can get toxoplasmosis from two sources: raw meat and cat feces.

Toxoplasma exists in many different kinds of animals throughout the world. It life-style varies with the host, a circumstance that contributes to the dual meat/cat droppings route of human infection. It is generally thought that *Toxoplasma* completes its development only in cats. The parasites live in cells of the cat's intestinal mucosa, where they undergo a typical, complicated sporozoan development. The results of all this development are oocysts that pass out of the cat in the animal's feces.

Other animals can become infected by swallowing the cysts, but when nonfelines swallow the oocysts, the development is quite different. In the nonfeline, various forms of the parasite can develop in practically any organ of the body, but the intestine is one of the least likely places. Just how much damage the parasite does depends on where it dwells and in what numbers, and how the host obtained the parasite. The parasite almost never develops to maturity in the intestine of nonfelines, but in cats, the parasites live mostly in the intestine and in other organs in lesser numbers.

Most people who accidentally swallow oocysts never develop symptoms. Some do, but the symptoms resemble those of so many other diseases that adult toxoplasmosis is seldom diagnosed as such. When there are symptoms they are much like mononucleosis. The patient is tired all the time, has headaches, and may have chills and fever. Occasionally the parasites get into the retina of the eye and cause problems there, though rarely so severe as to cause blindness.

Toxoplasmosis can also be obtained from eating raw or undercooked meat—usually beef, but the parasite can also be in pork and mutton. The parasites get into the cows, sheep, and pigs the same way they get into cats and people, that is, through the ingestion of oocysts. Cats, by the way, can get *Toxoplasma* from rat meat. When the animals eat the oo-

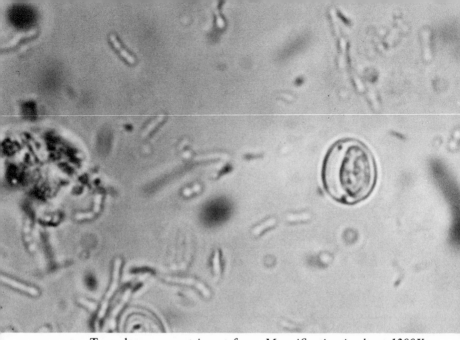

Toxoplasma *oocyst in cat feces. Magnification is about 1200X.*

YALE UNIVERSITY SCHOOL OF MEDICINE,
DEPARTMENT OF COMPARATIVE MEDICINE.
PHOTO BY ROBERT SULLIVAN

cysts, the parasites travel in the bloodstream until they reach various organs, including the muscles. The parasites survive digestion of the meat, go through developmental stages, and become established in the new host.

The incidence of toxoplasmosis is very high in countries where raw or just barely cooked meat is considered to be gourmet fare. France has one of the highest, if not the highest, rates of toxoplasmosis in the world. French children have a high incidence from being fed raw meat juice.

The most serious form of the disease is the congenital form, the form passed on to infants from their mothers before they are born. The parasites pass from the mother through the placenta into the developing fetus. The effects on the unborn child can be catastrophic. Symptoms include mental retardation, vision problems, and physical deformities.

The severity of the symptoms varies with the stage of development of the fetus when it is infected and whether or not the developing fetus has acquired any immunity from its mother. The estimates are that from 60 to 72 percent of the infants will have no symptoms at birth, but some may develop symptoms later. Ten to 13 percent will have some problems with vision, about 10 percent will have severe brain damage, and as many as 15 percent will die as infants. However, toxoplasmosis is usually limited to first-born children. The mother builds up antibodies during the first pregnancy that are passed on to the fetus during subsequent pregnancies.

Adults with symptoms can be treated. The drugs inhibit growth of the *Toxoplasma* organisms. Congenital toxoplasmosis usually does not respond well to treatment.

Transmission of the disease from mother to unborn child usually takes place during a new, acute, but undiagnosed infection in the mother. That is why the incidence is so high in countries such as France where the eating of undercooked meat is more frequent than most other places.

Congenital disease can be prevented. Pregnant women, especially those pregnant for the first time, should not eat raw or undercooked meat. If at all possible, they should avoid handling raw meat; *Toxoplasma* in meat has been known to get in through minute skin cuts. If raw meat has to be handled, wear rubber gloves. The hands should be washed thoroughly afterward. It is best not to have any cats in the house, especially if the cats are used as mousers. If there is a cat, some precautions are in order. Feed the cat dry or cooked canned foods, *never* raw meat. The litter box must be cleaned daily, and this cleaning must not be done by the expectant mother. Cat droppings should be flushed down the toilet and the litter pan disinfected with boiling water. Whoever does this unpleasant job must wash the hands thoroughly before coming into close contact with the expectant mother.

Acanthocephala

Acanthocephala is a phylum of worms that look like roundworms, but are really quite different. Each is equipped with a spine-covered proboscis at the anterior end, hence their common name, spiny-headed worm. They are probably more like tapeworms than roundworms, but they do warrant their own phylum in the opinion of most taxonomists.

They are mostly parasites of fishes and birds. However, there are two species that are sometimes found in human intestines. One, called *Macracanthorhynchus hirudinaceus*, is ordinarily a parasite of pigs. Intermediate hosts are grubs (larvae) of various beetles. The pigs get the worms by eating infected grubs. The worms develop into adults in the pigs' intestines, where the female worm can reach the impressive length of 60 centimeters (2 feet).

*Occasionally a few of these worms are found in people. Quite a few cases have been known to occur among peasants in southern Russia. Whether these infections were due to eating beetle grubs (there are many people who consider insect grubs to be a delicacy) or from associating too closely with pigs is not known. A heavy infection of these worms can be quite damaging. The spiny proboscis cuts and tears the intestinal mucosa, and sometimes perforates the intestine.

Dioctophyma renale

Earlier mention was made of a roundworm that is red rather than the usual white and is the largest-known parasitic roundworm. Well, this is the one. Called the great red kidney worm, it is ordinarily a parasite of mink, occasionally of dogs, and rarely of man. The female *Dioctophyma* can reach a length of about a meter (about three feet) and the thickness of a pinkie finger. As the name implies, it lives coiled up in the kidney of its host.

It is an important parasite of mink in the northern United States and Canada, infecting a fairly high percentage of ranch minks. Eventually the worm can kill the host by destroying its kidneys, although dogs have been known to live full life spans while harboring them.

The major concern is the possibility of accidental human infection. Fortunately such infections are rare. The life cycle is complicated, involving intermediate host stages in leeches and fish. The parasite gets into the main host when raw or undercooked fish containing the larval worms is eaten. One worm only is the usual infection. Generally the worm stays in one kidney until the organ is destroyed and then goes to the other one. The only treatment is to remove the affected kidney. The same precautions that apply to preventing fish tapeworm apply to the prevention of *Dioctophyma.*

Parasites of Pets and Domestic Animals

Pets and domestic animals are subject to getting hundreds of internal and external parasites. Most of them do their hosts very little, if any, harm, but some can make your pet pretty sick and others cause serious economic losses among livestock.

Diorfilaria immitis

Heartworm, *Diorfilaria immitis*, is a problem of great concern to dog owners. The parasite is similar to the filarial worms that infect people. The intermediate hosts are several species of mosquitoes. Microfilariae are found in the blood, but the main problem comes from where the adult worms live, which is the right ventricle (pumping chamber) of the heart and the nearby large blood vessels.

Like most filarial worms, these are long; the females are from 20 to 30 centimeters (8 to 12 inches) long. They can gather in a tangled mass in the dog's heart and cause the animal quite a bit of misery. Heavily infected dogs are generally short of breath. Periods of sudden collapse after exercise followed by prompt recovery is almost a sure sign of heartworm. Sometimes heavily infected dogs can drop dead quite suddenly.

Heartworm infection can be treated, but prevention is the better course. A medication called diethyl L-carbamazide given daily during the mosquito season prevents microfilariae from maturing into adults.

Humans can also get heartworm, but in people the development of the parasite is abnormal. A flat, coin-shaped growth develops. The usual diagnosis is "tumor" and only surgery or autopsy reveals the true nature of the growth.

Spiruroidea

There is a family of roundworms called Spiruroidea that are mostly parasites of fishes, but a few are economically important parasites of domestic animals such as ducks, geese, dogs, pigs, horses, and donkeys. Most of these have arthropod intermediate hosts. Two of them, *Habronema* and

Drascheia, are parasites of horses and donkeys. They cause these animals much gastrointestinal distress. One of the *Habronema* larval stages causes a condition called summer sores in horses. Other spiruoids are parasites of cattle, sheep, and pigs. These include the so-called stomach worms of pigs.

Gnathosoma are spiruoids that live in the stomachs of dogs, cats, and wild carnivores. These are damaging parasites that can drastically shorten the lives of their hosts. They are covered with spines, and as they burrow into the stomach wall they cause quite a bit of havoc. Although unable to establish in human hosts, they can cause a creeping eruption condition in people.

Flatworms

Flatworm parasites of pet and domestic animals are also numerous. A genus of tapeworm called *Moniezia* are parasites of cattle, sheep, and goats. The intermediate hosts are mites that nest on blades of grass, handily located for ingestion by the main hosts. The virtual invisibility of the intermediate hosts makes control of these worms practically impossible. Large numbers of these worms can accumulate in young animals, often causing intestinal obstruction. Animals seem to lose them as they mature.

Tapeworms

Horses are subject to invasion by a number of tapeworms, particularly those of the genus *Anoplocephala.* Some of these worms can have proglottids as wide as 50 millimeters (almost two inches), and have lengths of about a meter (three feet). Mites that gather on blades of grass are the intermediate hosts. Large numbers of the worms can irritate the intestine

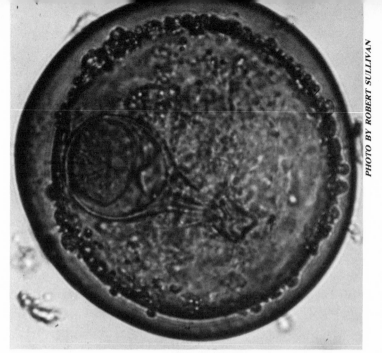

PHOTO BY ROBERT SULLIVAN

Egg of Moniezia, *the large tapeworm of sheep and cattle.*

YALE UNIVERSITY SCHOOL OF MEDICINE,
DEPARTMENT OF COMPARATIVE MEDICINE.

and cause intestinal blockage. Treatment is effective, but prevention is almost impossible since the mites are just about everywhere. Treatment is really the most effective prevention; more treatment means fewer worm eggs for the mites to ingest. Moving horses to new pastures never before, or not recently used, by horses is an expensive and only partially effective preventive action. Feeding horses only on bagged feed grains is another expensive preventive measure. However, it is difficult to keep the most regally fed horse from nibbling at grass, and there might be a few infected mites in the feed grain.

Dogs that live in rural and suburban areas where they might be able to catch a rabbit every now and then might pick up a tapeworm called *Taenia pisiformis* (also called *Taenia serrata*) as a prize for their victory over the rabbit. Cats can also get this tapeworm. The worms exist in larval stages in rabbit meat in much the same way larval stages of human tapeworms exist in beef and pork.

Domestic birds are plagued by a variety of tapeworms. The intermediate hosts of water birds are aquatic animals such as cyclops and snails. Tapeworms of ground birds such as chickens utilize insects as intermediate hosts, although slugs (shell-less snails) are intermediate hosts of at least one chicken tapeworm. They can make the birds quite ill, and they always slow down weight gain, reducing the profits of the poultry growers.

Protozoa

There are thousands of protozoan species that abound in pets and domestic animals in relationships ranging from mutualism to death-dealing parasitism. *Coccidia*, sporozoans similar to *Toxoplasma*, are parasites of poultry and all kinds of livestock including both furry animals such as rabbits and hoofed ones such as cattle. The species, however, are particular about their hosts; a species that infects pigs, for example, will not do well in cattle. Turkeys are subject to infection by a particularly damaging species. Often entire flocks of turkeys (and other birds) are wiped out practically overnight by heavy infections of *Coccidia.* A few species, by the way, can infect humans; symptoms of such infections vary as widely as do those of *Toxoplasma.* Expensive measures are needed to reduce the incidence of *Coccidia* infection among flocks of domestic birds. These costs are passed on to everybody in the cost of chickens and turkeys.

Arthropods

Arthropod parasites cause serious economic losses. Hoofed mammals such as cows and horses are bothered by flies generally referred to as "bots." These flies deposit their eggs on various parts of the animal's body. The maggots that hatch out go through various stages of development in the host and then exit by a variety of means.

Horses are bothered by bots of the genus *Gasterophilus*. These are variously referred to as lip bots, nose bots, chin flies, and throat bots. The flies lay their eggs on the hairs of the horse, generally around the mouth. Although these flies do not bite or sting, they annoy the horse mightily as they lay their eggs. The horse licks the eggs into its body. Some bots develop in the stomach and drop out in the feces as pupae. Others develop in the mouth and nasal passages. They can damage the stomach, intestines, and mouth and nasal regions of the horse. Treatment is available, and prevention can be accomplished by washing the horse with medications that will either kill the eggs or cause them to hatch out prematurely and then kill them. Mechanical devices that keep the flies away from the horse's mouth are also used.

Flies of the genus *Hyposerma* are called bomb flies and heel flies. These flies lay their eggs on cattle. The eggs are usually laid on the legs. The maggots migrate to the walls of the stomach and esophagus, where they stay for a few months. Then they move up toward the back of the cow, eventually reaching areas near the skin surface, where they complete their larval development in little pockets in the cow's back. When the maggots are fully grown, they drop off and pupate in the ground. The northern bot, or bomb fly, frightens the cow with the noise it makes. Often the cows seem to go berserk, rolling around and running wildly, often injuring themselves. Both control and treatment are difficult.

Dogs are sometimes parasitized by an arthropod that looks more like a worm than an arthropod. These are the so-called tongue worms, *Lingulata serrata.* They live in the nasal passages and frontal sinuses of dogs, where they subsist on blood. Symptoms can include sneezing, bleeding, and nasal discharge, or there may be no symptoms at all. Occasional human infections have been recorded. The life cycle is complicated, involving stages in sheep. The worms are fairly large for the small abode in which they live; females attain a length of about five inches.

Like the effort to control insects, the battle to control parasites in pets and domestic animals is constant and expensive. Just as growing one species of plant in a mass called a crop has encouraged insects, so has the domestication of animals exposed them to greater danger of infection by certain parasites. It is up to their human masters to control these parasites. Otherwise the animals would soon be overwhelmed by them.

If you have a pet, you have the responsibility of having the animal examined for parasites and treated, if necessary, at least once a year. That is as much an obligation as feeding the animal. Keeping the animal free of parasites is also important to the health of you and and your family.

10.

Constant Problems,
Often Ignored

After reading this book you might very well know more about many of the parasites discussed here than some American physicians. That is no exaggeration. I have come across many physicians who know nothing or very little about schistosomiasis, filarial worms, and many other human parasites. Such lack of knowledge does not mean they are bad doctors, but that American medical schools tend to give very little attention to these parasites in the course of educating the nation's physicians.

The reason for this neglect is that parasites such as tapeworms, roundworms, and flukes are not considered to be a great health problem in the United States. In some respects that assessment is correct. The incidence of these parasites in the United States has gone down considerably in recent years; many medical educators and physicians do not consider them to be worth a great deal of attention. There are so many other problems occupying the medical attention of the nation: cancer, heart disease, arthritis, and mental diseases, to name just a few.

Although most American physicians know of the existence of tapeworms, roundworms, and such, the possibility that these organisms might be the cause of a patient's problem seldom enters the diagnostic thinking of many American physicians. Often the wrong treatment is used, treatment that leaves the parasites unscathed but leaves the patient a little poorer in time, money, and strength. In many cases parasites continue to thrive, shortening the life of the patient. Congenital toxoplasmosis is but one example of the medical tragedies that could be prevented with just a little more thought to the possibility of parasitic infections.

Diagnostic tests of various kinds have been a big help in finding unsuspected parasitic infections. A routine X ray often reveals a suspicious-looking mass that turns out to be a hydatid cyst rather than a cancer. Barium X ray procedures (GI series) sometimes show that the suspected ulcer or irritable bowel syndrome is really a bountiful crop of ascarids.

American physicians might encounter parasites with greater frequency in the near future. Fast transportation continues to make the world smaller, and travel has always helped to spread parasites. The influx of refugees from areas such as Southeast Asia is bringing the problem of parasitism home to American physicians, many of whom are seeing afflictions such as schistosomiasis, guinea worms, and lung flukes for the first time in their careers. There are thousands of people with diseases such as these in large cities like New York.

Although the possibility of the life cycles of these parasites becoming established in the United States is remote, it is definitely a possibility and is one that requires constant vigilance. Recall that there was no hookworm, schistosomiasis, or onchocerciasis in the Western Hemisphere until the slave trade brought them here.

Quite often when a diagnosis of an internal parasite infection is made, the physician finds there is no medication available with which to treat it. Local drug suppliers are not likely to stock antischistosome medications, for example, and many American pharmaceutical manufacturers don't bother with them because the demand is not great enough to make them profitable. Sometimes it is necessary to go to foreign sources to obtain the drug, and that usually requires special permission from the Food and Drug Administration if the drug is not approved for use in the United States. However, the Center for Disease Control in Atlanta maintains stocks of many antiparasitic drugs.

The parasites discussed in this book may be relatively rare medical problems in the United States, but they are not about to become extinct. They will always be here, waiting in their own unthinking but effective way to take advantage of every opportunity. We are supposed to be the thinking animals, but it has always been the stupidity of humanity that has given these parasites their big chance. The Black Plague followed in the wake of the wars and strife that were so much a part of everyday life in the fourteenth century. Malaria swept across southern Europe as the Roman Empire crumbled, and schistosomiasis became the scourge of Egypt when that nation sank into poverty and decline after successive waves of conquering armies swept across the country. There were typhus epidemics in Europe during and after World Wars I and II, and in Korea during that country's wars. Trichinosis broke out in epidemic proportions in Poland after World War II when there was no governmental structure to control the manufacture and marketing of pork.

It can all happen again. A relaxation of standards of sanitation such as allowing the dumping of raw sewage into lakes and rivers could fill many of us with tapeworms, ascarids, and a host of other guests. Local upheavals have already resulted in some small outbreaks of plague in Africa. Another world war could bring back massive plagues and worm infections. They're waiting out there, and they have plenty of time. Just how successful these parasites will be at our expense is up to us, not them.

Glossary

Adaptations: All of an animal's characteristics that help it survive.

Alveolar cyst: A free-form-shaped hydatid cyst.

Autotroph: An organism able to synthesize organic food directly from inorganic material.

Bot: A type of fly parasitic in hoofed mammals.

Brood capsule: A globular mass of cells produced by a hydatid cyst.

Carnivore: A member of the order Carnivora, a group of flesh-eating mammals.

Cercaria: An infective larva produced by digenetic trematodes, stages in the intermediate host.

Commensalism: A living arrangement in which one partner benefits and the other is not harmed or benefited.

Coracidium: The swimming, first larval form of the fish tapeworm.

Cyst: A protective capsule that surrounds larval stages of some parasites.

Cysticercus: A larval form of the tapeworm.

Ectoparasite: A parasite that lives on the host.

Endoparasite: A parasite that lives in the host.

Facultative parasite: An organism that can live alone, but can also parasitize under the right conditions.

Gametocyte: In plasmodia, the sexual forms that are ingested by mosquitoes to complete the life cycle.

Genes: The basic units of heredity.

Herbivore: A plant-eating animal.

Hermaphrodite: An animal with both male and female reproductive organs.

Heterotroph: An organism that cannot manufacture its own food and must therefore eat plants or animals to live.

Host: The organism the parasite lives on or in.

Hydatid cyst: Larval stage of *Echinococcus granulosus.*

Hyperparasite: A parasite that lives in or on another parasite.

Infection: An invasion of an organism by parasites that reproduce inside the host's body; in this book usually used for parasites such as worms.

Infestation: An invasion of an organism by parasites that reproduce outside the host's body.

Instinct: The preprogrammed responses an animal inherits.

Intermediate host: A host used before the parasite becomes established in the primary host.

Life cycle: The series of events in an organism's life, from birth to reproduction to death.

Main host: *See* primary host.

Merozoite: The result of *Plasmodium* sporozoite reproduction in red blood cells.

Metacercaria: The final larval form of the liver fluke.

Microfilariae: The young of filarial worms.

Microgamete: The mature *Plasmodium* gametocyte.

Miracidium: A free-swimming larval form of digenetic trematodes.

Mutation: A spontaneous change in the gene.

Mutualism: A living arrangement in which neither partner can live without the other.

Myiasis: Maggot infestation.

Obligatory parasite: This organism must be a parasite to live.

Omnivore: An organism that is both plant- and animal-eating.

Oökinte: An active *Plasmodium* zygote.

Pathogenic: Disease-causing.

Periodic parasite: A parasite that visits a host only occasionally.

Phoresis: A living arrangement in which one animal is physically carried by the other; of doubtful or little benefit to either.

Plerocercoid (or sparganum): The third larval form of the fish tapeworm, found in freshwater fish.

Predator: An animal that lives by killing and eating other animals.

Primary host: The basic host in a parasite's life cycle.

Procercoid: The second larval form in the fish tapeworm, found in tiny freshwater organisms.

Proglottid: A segment of a tapeworm.

Redia: The result of trematode sporocyst reproduction.

Saprophyte: An organism that lives on the remains of dead organisms.

Scolex: The "head" of a tapeworm.

Spargana: *See* Plerocercoid.

Sporadic parasite: *See* Periodic parasite.

Sporocyst: The saclike structure into which a trematode miracidium changes once inside an intermediate host.

Sporozoite: An infectious *Plasmodium* stage.

Symbiosis: A close living arrangement.

Symbiot or symbiont: An organism living in symbiosis.

Taxonomy: The science of classification.

Trophozoite: The active stage of *Entameba histolytica.*

Vector: A carrier of a parasite.

Zygote: The product of the fusion of a male and a female microgamete.

Bibliography

Beck, Walter J. and J. Davis, *Medical Parasitology*. St. Louis, Mo., C. V. Mosby, 1976.

Brown, H. W., *Basic Clinical Parasitology*. New York, Appleton-Century-Crofts, 1971.

Chandler, Asa C. and C. P. Read, *Introduction to Parasitology with Special Reference to the Parasites of Man*. New York, John Wiley and Sons, 1961.

Cheng, Thomas C., *General Parasitology*. New York, Academic Press, 1973.

Jones, A. W., *Introduction to Parasitology*. Palo Alto, Cal., Addison-Wesley, 1967.

Kennedy, C. K., ed., *Ecological Aspects of Parasitology*. New York, Elesevier North-Holland, 1976.

Markell, Edward K. and Marietta Voge, *Medical Parasitology*. Philadelphia, Saunders, 1976.

Meyer, Martin C. and O. W. Olson, *The Biology of Animal Parasites*. Philadelphia, Lea and Febiger, 1976.

Noble, Elmer R. and Glenn A. Noble, *Parasitology: The Biology of Animal Parasites*. Philadelphia, Lea and Febiger, 1976.

Read, Clark P., *Animal Parasitism*. Englewood Cliffs, N. J., Prentice-Hall, 1972.

———, *Parasitology and Symbiology: An Introductory Text*. New York, John Wiley and Sons, 1970.

Schmidt, Gerald D. and Larry S. Roberts, *Foundations of Parasitology*. Philadelphia, C. V. Mosby, 1977.

Sousby, E. J., *Helminths, Arthropods, and Protozoa of Domesticated Animals*. Philadelphia, Lea and Febiger, 1968.

Taylor, Angela E. and Ralph Muller, eds., *The Relevance of Parasitology to Human Welfare Today* (Symposia of the British Society for Parasitology: Vol. 16). Philadelphia, Lippincott, 1978.

Index